A Chanticleer Press Edition

Lorus and Margery Milne

THE ARENA OF LIFE
The Dynamics of Ecology

Doubleday/Natural History Press
Garden City, New York

*Published in the United States by Doubleday & Co., Inc.,
New York*

*Planned, prepared, and produced by Chanticleer Press,
New York*

Manufactured by Amilcare Pizzi S.p.A. in Milan, Italy

Library of Congress Number 71–159519

Contents

For Minnie Mae and Don Murray
as they share the living world
with their own contributions to it.

Foreword

On each continent, plants and animals follow their own distinctive ways. They meet the challenges of life on earth. Today, as never before, man looks intently at these living things with which he shares the only livable environment he can reach. He recognizes that many kinds of animals and plants no longer range as widely as they did a century ago. He sees that his own competition with nonhuman life, particularly through intensive cultivation of a few favored species, has transformed forests into fields, and fields into deserts. Despite improved methods for harvesting the seas, the oceans yield less food annually. Two-thirds of mankind barely subsists. Yet the human population soars as though not subject to the ecological principles that govern all other forms of life.

The impact of human activities on the ecology of the world is scarcely appreciated even in the nations with the highest level of education. In less industrialized countries, customs in diet and in family planning lag still further behind modern understanding of human needs. Few people have yet absorbed the realization that all living things are now imperilled by the struggle to feed the increasing human population.

In the last three centuries mankind has subsidized technology with energy from sources that are renewed only through geological processes requiring millions of years. This cultural development, based upon exploiting resources faster than they can be recycled, has altered the conditions for life on earth through spectacular accumulations of nonliving wastes and living people. It has made our species increasingly dependent upon maintaining an artificial environment, and prolonged the accep-

tance of a past misapprehension. The world does not belong to man to do with as he chooses. Mankind belongs to the world, and can have a future only as a cooperating part of its ecology.

The maintenance of a suitable environment for life depends on a chain of energy and on balances in nature that have taken billions of years to evolve. To understand the ecological principles in this system, we have investigated the ways in which plants and animals meet the challenges in the oceans, in fresh waters and the soil, in rain forests, in grasslands where drought is seasonal, in deserts, at high elevations and high latitudes where intense cold prevails. In all of these situations we have looked for ways in which mankind might quickly minimize the impact of technology upon the environment. To keep man's options open, a prompt change from past procedures seems essential.

We have benefitted greatly from discussions with experts who have been generous with their time and knowledge. Our heartfelt thanks go to the many people who have helped us become acquainted at first hand with plants and animals on six continents, and to scientists who have made available in scientific journals their discoveries about the world. We are grateful particularly to Dr. Milton Rugoff of Chanticleer Press for his advice and encouragement, and to his editorial associate Miss Jean Walker, and to Mr. Ulrich Ruchti and Miss Elaine Jones for their resourcefulness in collecting and presenting the illustrations. Hovering helpfully behind the scenes has been Mr. Paul Steiner of Chanticleer Press.

April 28, 1971   *L.J.M. and M.M.*

1 Beginnings of ecology

Each kind of living thing thrives in some sites but not in others. Usually it is found where conditions are suitable for its reproduction as well as its survival. People have realized this in a general way for many millennia. Yet, until the past century, no systematic scientific study was made of the interrelationships of plants, animals and the physical environment. This study constitutes ecology.

Long before anyone looked for the underlying relationships, people learned how to influence the food productivity of an area. The American Indians, who hunted bison on the Great Plains, set fire to the tall grasses near the border of the eastern forests. The fire spread among the trees, killing them and making a place for more short grass and bison. In the Orient, people living on mountainsides and on the flood plains of the great rivers discovered how to hold the water in diked paddy fields. By diverting the natural runoff, they could raise edible crops of rice and taro where these plants would not ordinarily grow.

The association of plants with particular areas was pointed out by Theophrastus of Eresus, who studied with Aristotle and was responsible for the earliest known botanic garden. In his *Historia Plantarum*, written before 325 B.C., he comments that plants

> all are distinguishable as either terrestrial or aquatic, just as we also primarily distinguish animals; for there are some plants which grow nowhere but in the sea. . . . Others affect only marshes or other very wet places. Some cannot live in very wet ground, but restrict themselves to dry ground. Certain others are littoral only.

This awareness, developed in classical Greece, dwindled during Roman times and almost disappeared during the Middle Ages. It received new stimulation when learned men accompanied the Crusaders and returned to Europe after seeing with their own eyes that the flora and fauna of Asia Minor differed from those of central and western Europe. The descriptions in the writings of Aristotle and Theophrastus applied to the living things of countries around the eastern Mediterranean, but not with the living things in Switzerland, Germany, and the British Isles. The travelers realized for the first time that living things were far from uniformly distributed around the world. Local plants and animals needed separate examination.

A number of physicians in Europe began in the 16th century to assemble all available information and to incorporate it in encyclopedic books known as herbals (accounts of herbs and other plants) and bestiaries (about beasts). One of the most able of these men was Jerome Bock, who in 1539 published *Das Neue Kreuterbuch*, in which he described the kinds of plants he knew from the woods and fields of his native Bavaria. He mentioned the type of location in which each plant grew. The five-volume *Historia Animalium* by Konrad Gesner of Zürich was far more ambitious. It was the first illustrated account of the animals of the world and included information on the geographical distribution of non-European species.

As explorers continued to return with a vast amount of information from distant lands, the need for a manageable classification grew acute. The invention of the microscope and the discovery of an unsuspected living world in miniature also increased the urgency. In 1758 a patient Swedish physician-naturalist, Carl von Linné, or Linnaeus, devised the first con-

sistent method of giving scientific names to living things and his system has been in use ever since. Biologists journeyed far to find species new to science. Generally, the notes they sent to museums with the specimens found a place in the records, but often had to be simplified.

Today we can appreciate how little the pioneers knew and what hardships they underwent in their explorations. They traveled where they could: George E. Rumpf to the Dutch East Indies, Hans Sloane to the British West Indies, Mark Catesby to Virginia, Florida, and the Bahamas, and John Bartram from Ontario to Florida, seeking plants for his botanic garden in Philadelphia. Daniel Solander, a pupil of von Linné, accompanied Joseph Banks as the naturalist aboard the British barque *Endeavour*, which circuited the world under the command of Captain Cook between 1768 and 1771. Carl P. Thunberg stopped at South Africa and then went on to Japan at a time when only Dutch

travelers were admitted there. Alexander von Humboldt used every available form of transportation through South America from the Venezuelan coast to the Andes and back. Returning to Germany in 1804, he published twenty-six volumes based on his five years in tropical and temperate America. His voluminous records include the air temperatures, altitudes, relative abundance of plants and their associations in the areas where he found his specimens.

The thoroughness of von Humboldt inspired many later explorers. Charles Darwin felt almost obligated to do as well during his years (1831–1836) aboard H.M.S. *Beagle* while it was circling the world. Darwin's friend, Joseph Hooker, a botanist as well as a surgeon, went on an expedition led by Sir James Ross to the antarctic regions (1839–1843). Alfred R. Wallace and his friend Henry W. Bates set out for the Amazon in 1848, hoping to finance their travels by selling specimens to British and continental museums. Disagreeing with Bates on many matters, Wallace returned a year later to England, but soon was off again—to Malaya and the East Indies. His experiences in the eastern tropics led him to write three books that were to prove very influential: *The Malay Archipelago* (1869), *The Geographical Distribution of Animals* (1876), and *Island Life* (1880). Although Wallace is usually remembered along with Charles Darwin as a co-discoverer of the principle of natural selection, he is respected at least as much for filling various gaps in the knowledge of the geography of life. In this area he progressed far beyond P. L. Sclater, who tried to subdivide the world according to the birds found on each continent.

Despite the fact that most plants have a fixed habitation, the botanists needed much time to find the underlying features relating vegetation to climate and other aspects of the environment. The role of rainfall and temperature throughout the year impressed Alphonse de Candolle. In 1874 he pointed to fundamental differences among the plants of the hot wet tropics as compared with those of hot deserts, temperate forest areas, savannas and grasslands, and tundras. Extending this analysis, the German botanist A. F. W. Schimper investigated the distinctive features of different regions and summarized them in a book published in 1898. Translated into English as *Plant Geography upon a Physiological Basis*, the volume became a classic. The American ecologist Victor E. Shelford refined the conclusions into the "law of the optimum" (1913), in which he suggested that any species of plant or animal achieves success where its tolerances are put under least stress by the factors in the environment.

Plant geography and natural history promised to become more exact through study of the environmental features that affect the suitability of each place for particular plants and animals. In 1868, the German biologist Karl Reiter gave this new branch of biology the name Oekologie, from the Greek *oikos* (a home) and *logos* (a discussion). The following year, his countryman, Ernst H. Haeckel publicized the word. In English, it became ecology.

Paradoxically, the conclusions reached by Charles Darwin and Alfred Russel Wallace, who were experienced field naturalists, sent contemporary biologists indoors to seek evidences of evolution from preserved specimens. Even the establishment of oceanography as an important field for study, through the pioneering investigations of scientists aboard the research vessel H.M.S. *Challenger* (1872–1876), did scarcely more for ecology than add to museum collections a host of peculiar animals from the depths of the oceans. Variations in living things according to depth and temperature in this realm, which covers more than two-thirds of the globe, took longer to appreciate.

Slowly the number of ecologists increased. Those in England founded a British Ecological Society in 1912 and began publishing the *Journal of Ecology*. The American scientists who had cooperated in managing *Plant World* decided to broaden and transform their enterprise into the periodical *Ecology*, which was begun in 1920. The newly founded Ecological Society of America managed the publication. For the purpose of publishing longer reports, the society created *Ecological Monographs* in 1930.

For nearly twenty years, ecologists wrestled with two major handicaps in their attempts to understand living things in the natural environment. Each individual plant or animal is subject to so many different influences in a single day, or growing season, or lifetime, that measurable changes can rarely be matched to definite causes. In addition, the samples that an ecologist can collect are often hard to identify: the food of an animal from stomach contents, fragmentary and partly digested, or the fungi associated with a plant, without any distinctive reproductive structures. Years went by while ecologists tracked down the facts they needed, and learned to pose scientific questions for which useful answers could be reached. During those years, the tools of the biologist improved spectacularly. Electronic computers facilitated the analysis of changes due to multiple factors. Radioactive tracers, autoradiography, various types of chromatographic analysis, and other delicate measuring techniques all came into use.

At first, ecologists analyzed environments and interactions among nonhuman life and tried to ignore the presence and effect of mankind. Their reports read as though human beings did not exist. Wilderness for study was still available. And with no general awareness of the devastating influence of mankind, support could scarcely be found for restricting human progress. It seemed better to ignore the pessimistic conclusions of Charles Perkins Marsh, whose book *Man and Nature* (1864) pointed to so many places where exploitation of the land had led to ruin.

Following World War II, thoughtful men came to the realization that man's impact on nature had become nearly universal, that man determined to a large degree what plants and animals survived and the consequent effect on human welfare.

11

In his Pflanzen-Geographie, *A. F. W. Schimper, of Basel, noted that the common yarrow* (Achillea millefolium) *grows alongside the slightly different* A. moschata *(left) and* A. atrata *(right) but the latter two rarely together, since they make similar demands on the environment.* A. moschata *excludes* A. atrata *on silicious soil, and* A. atrata *excludes* A. moschata *from calcareous soil.*

Human ecology became a frontier, if only to find space, food and other resources for an exploding world population. Ecologists are now attempting to combine their findings with the thinking of economists and sociologists to establish priorities for action.

The need for more information and for fundamental theories in ecology that could be applied to human problems has led biologists to seek international scientific cooperation on a grand scale. They have organized an International Biological Program (IBP), which began in 1967 and will continue for a number of years. Special efforts are being directed toward standardizing techniques of measurement so that findings on one continent can be compared with those on another, and information regarding the productivity of life in terrestrial situations can be compared with that in aquatic environments. To use scientific resources efficiently during the IBP, worldwide research is being concentrated on a small number of topics: convergent and divergent evolution, analysis of ecosystems, the physiology of colonizing species, the biogeography of the sea, phenology (the comparative study of the timing of seasonal events), and the conservation of environments. Five IBP programs are focusing on human adaptability, namely, the adaptations of Eskimo populations, population genetics of the American Indian, the ecology of migrant peoples, the biology of human populations at high altitudes, and nutritional adaptations to the environment. All these studies are expected to provide the scientific basis for international cooperation aimed at ensuring "man's survival in a changing world."

Today, ecologists investigate patterns of interactions at levels beyond that of the individual or even the species. They study population, community, and ecosystem ecology on the one hand and physiological and behavioral ecology on the other. In all of these fields the ecologist has become intent on following the flow of energy through living organisms and on tracing the cyclic movement of chemical elements through the ecosphere because it is this energy and these chemical materials that nourish mankind and all other living things.

Basically, the ecologist seeks to comprehend the interaction between inherited "nature" and environmental "nurture." The inherited nature sets limits to body structure and function. These limitations are crucial to each living thing as it meets the challenges in its environment and as it competes for the available resources. Inheritance largely determines whether a species will get its energy directly from the sun, or by eating plants or meat, or through a mixed diet, or by decomposition or some parasitic habit. The source of energy used by each individual often indicates the dangers to which it is vulnerable. The genetic pattern also specifies whether the species will go through two generations in an hour, as some bacteria do, or four generations in a century, as is usual in man.

The behavioral ecologist studies the ways in which different animals get their energy and the chemical materials for growth and reproduction. Some individuals, he finds, can ignore one another, whereas others are clearly affected by having neighbors close by. Within a species, where competition for environmental resources is most acute, behavioral differences can become adaptive and diminish competition. They can lead to territorial subdivision, to social cooperation, or to new ways of life that may eventually result in the development of new races, subspecies, and even species.

The population ecologist follows the fluctuations in the numbers of individuals of a species, perhaps by age groups, and tries to separate the effects of chance variations in environmental features, such as climate, from variations related more intimately to the physiological characteristics of the specific plant or animal.

The ecosystem ecologist attempts the broadest view. He studies the interactions and adaptations of life in areas of the world (the "ecosphere") that are so large that events in other divisions can be ignored. The division, or ecosystem, may be an ocean, soil, forest, grassland, or desert. He tries to account for all of the energy that enters the ecosystem, that is stored or transferred, and that eventually is lost. He measures the productivity and efficiency of transfer as energy goes from one form of life to another.

The community ecologist focuses on the ways in which the members of neighboring species interact. He measures the ways in which the environment of each living thing is altered by the presence of other living things, causing the whole community to undergo a slow succession of changes.

Various environments reveal features related to the evolution both of continents and islands and of life on earth. In this area the efforts of the ecologist merge with those of the biogeographer to account for the present distribution of living things. They contribute to our comprehension of the world as the human species explored and colonized it, exploited and altered it. They give us the perspective needed to appreciate the explosive rate of man-made change and for cherishing the only ecosphere we have.

2 Energy for life

Incredible as it seems, life on earth exists only because of a chemical reaction in the sun, some 93 million miles away. It is a nuclear-fusion reaction of immense magnitude, virtually identical with that of a hydrogen bomb. Like a thermonuclear reaction, it occurs only at high temperatures, but instead of ending in one tremendous explosion, the process in the sun is continuous. Steadily it converts the matter of solar hydrogen into helium and the radiant energy of sunlight.

The energy from the sun radiates in all directions, decreasing in intensity as it goes. At the average distance of the earth from the sun, the radiation is still effective although the huge ball of the sun subtends only a small angle—approximately two minutes of arc. If we could look from the sun toward the earth, our planet (whose diameter is a mere 1/109 as great as that of the sun) would appear only as a dot, scarcely more than one second of angle from side to side. Consequently, the earth intercepts just about one-535-billionth part of the energy radiated by the sun. This insignificant fraction sustains the only life we know.

At any distance from the sun, a planet captures radiant energy and then reradiates the same amount as heat to outer space. This equilibrium holds the temperature of the earth's surface in the narrow range at which water can be liquid while additional amounts circulate in gaseous form. The dimensions of the earth make the gravitational force hold water vapor and other gases in an atmosphere. The earth, moreover, rotates fast enough to redistribute the atmosphere, and thus prevents the development of temperatures too high or low for life to tolerate.

At the outer boundary of the earth's atmosphere, the solar energy arriving each second amounts to 1.95 gram-calories per square centimeter or 0.013 kilocalories per square inch. Known as the solar constant, this measure embraces a wide range of wavelengths that are measured in millimicrons or nanometers, each a billionth (10^{-9}) of a meter. Wavelengths increase from cosmic rays, X-rays, and ultraviolet shorter than 400 nanometers, through the visible violet, blue, green, yellow, orange, and red, into the longwave zone beyond 700, which is the infrared.

At sea level we hardly realize how much of the sun's energy has been selectively absorbed by clear air. A stratum of ozone gas at an altitude of 14 to 16 miles (22 to 25 kilometers) is virtually opaque to wavelengths from 290 down, and still shows important effect to 320 millimicrons. Along with oxygen, it converts the absorbed energy into heat and prevents it from reaching living things. Since radiations shorter than 300 millimicrons have strong antibiotic action, through disruption of organic molecules, the ozone screen actually makes terrestrial life possible. Wavelength 300 still "sunburns" human skin.

The infrared portion of the solar spectrum gives sunlight much of the warmth we feel. It also contains about half of the solar energy that reaches the outer boundary of the atmosphere. Between that and sea level, however, the wavelengths from 3,200 to 2,300 millimicrons are filtered out by carbon dioxide, with some help from ozone and water vapor. Those from 2,300 to 720 have a chance to get through if they are not intercepted by water vapor or by clouds of condensed moisture or ice crystals, which are even more effective. All of the radiant energy that is absorbed in the upper atmosphere contributes to the power of the winds that redistribute heat between the Equator and the Poles.

In combination, these same gases produce the phenomenon known as the "greenhouse effect." They capture and convert to heat much of the outbound infrared waves that are emitted by areas of the planet warmed by sunlight. By preventing the escape of such energy to outer space, the gases keep the earth's temperature significantly higher. This is why a thick cover of low clouds at night provides a warm blanket, whereas under a clear sky the thermometer shows a marked drop. Still more dramatic is the radiational cooling between sunset and sunrise over arid lands, where water vapor in the atmosphere is so scarce that the reradiated infrared escapes with great rapidity.

The portion of the spectrum we see as red and orange is affected negligibly by molecules of air and fine dust particles. These components of the atmosphere scatter the shorter wavelengths and give us the impression that the sky is blue. The scattering effect increases geometrically toward the violet end of the spectrum and prevents about half of the energy there from reaching the earth directly even when the sun is overhead. While the sun is near the horizon, as before sunset or after sunrise or at noon in a high latitude, the light passes obliquely through the atmosphere. It loses its violet and blue while the longwave red comes through disproportionately. Correspondingly, the light at sea level is dominated by the red, and "white" light then becomes pinkish.

The solar spectrum narrows even more quickly whenever the energy penetrates into aquatic environments. Water is almost opaque to infrared. Virtually all of the red is absorbed by the topmost 10 feet and the orange by the topmost 25 feet. Below, all red animals and plants appear black to a diver. Few green algae live below this level. Yellow disappears 50 feet down, but brown algae use the narrowed spectrum and lesser intensities down to 100 feet or more. Below that only the red algae carry on photosynthesis and, at more than 350 feet below the surface, they too starve for light. At a depth of 600 feet, only the blue and blue-green are left and the total energy from sunlight has diminished to less than a hundredth of one per cent. And finally, a narrow band centered around 475 millimicrons in the blue passes the 600-foot level.

Light of Life

Solar energy penetrates the atmosphere at low altitudes in a continuous spectrum of wavelengths from 340 to 1100 millimicrons and becomes available to living things, which use the energy for many different photochemical reactions. Of these, photosynthesis is of paramount significance for life on earth because it provides the wealth of organic compounds that nourishes green plants, nongreen plants, animals, and all of the agents of decay as well. Photosynthesis is second in importance for life only to the production of radiant energy in the sun.

The absorption of solar energy for photosynthesis cannot take place without pigments called chlorophylls. They can utilize any of the wavelengths available to them at sea level, but do so most efficiently around 440 in the violet and 660 in the red. In white light, they transmit and reflect disproportionately in the yellow-green to blue sector, and thereby show the green color our eyes detect.

17

A fourth of the solar energy that penetrates the earth's atmosphere is absorbed in the evaporation of water, sustaining the great hydrologic cycle. Green plants in the sea capture a smaller fraction and become most efficient when the intensity of sunlight is about a tenth of its maximum at sea level (left and below) as when shafts of sunlight pass through gaps in the cloud cover and the sun itself is low in the sky. Lightning discharged between clouds (bottom right) or between cloud and earth can synthesize ammonia as a fertilizing component of rain.

The thermonuclear reaction in the sun sustains a temperature of 10,963°F (6,073°C), causing immense swirls of incandescent gas to rise from sunspot areas. These changes alter the character and intensity of the radiant energy emitted in all directions, and hence also the conditions of life on earth, some 92 million miles away.

Although we recognize the importance of green plants in aquatic environments, we rarely realize how poorly suited their chlorophylls are for capturing the solar energy that passes more than a few feet into water. At moderate depths the spectral range is so narrow that the plants can carry on photosynthesis only through use of accessory pigments. These absorb energy and transfer chemical excitation to the chlorophylls. In trying to comprehend why chlorophylls still have a place in these reactions, the distinguished biochemist George Wald looked for other properties that might single out the chlorophylls and found a unique inertness that allows them to store captured energy momentarily without harm. They release the energy smoothly for the photosynthetic step in which hydrogen, captured from water molecules that have been split into hydrogen and oxygen, is transferred to join atoms from carbon dioxide in producing the organic compounds.

Whether as sunlight and skylight by day, or as moonlight and skylight at night, most of the illumination reaching sea level comes in the range from 400 to 700 millimicrons. This is the part of the spectrum to which our eyes are sensitive. From intensities of full moonlight upward, the cone cells of our retinas give us color vision; they show maximum absorption in the greenish–yellow around 555 millimicrons. With less illumination we rely, as other vertebrates do, upon our rod cells; they are less sensitive to red and more to blue, with a peak of sensitivity around 520 in the green.

Visual systems of animals living far down in the sea have their greatest sensitivity in the band of wavelengths penetrating the farthest. On land, by contrast, many of the insects that are active by day see best in ultraviolet light beyond the absorptive range of the chlorophylls in green plants, with a peak in sensitivity around 365 millimicrons. For them the ultraviolet reflected by a flower is far brighter than any color we see. A honeybee sees ultraviolet as a separate hue. Although it has color vision, it is completely blind to red.

Green Traps for Solar Energy: The Producers

For at least half a billion years, green plants in the upper levels of the sea have been living by photochemistry. In addition to solar energy, they need water as a source of hydrogen, carbon dioxide for both its carbon and oxygen, and small amounts of inorganic substances such as phosphates, nitrates, potash, and lime. As long as these are available, the green plants are almost independent of other kinds of life. They are self-nourishing producers and hence called autotrophs from the Greek *auto* meaning self, and *trophein* meaning to nourish. In the ability to synthesize organic compounds with energy absorbed from sunlight, green plants are almost alone. Fortunately for nongreen plants, for animals and mankind, all of which live at the expense of the green plants either directly or indirectly, the producers carry on far more photosynthesis that is necessary to meet their own needs. They provide a substantial surplus, which not only nourishes every other living thing but often accumulates as dead wood, peat, coal, natural gas and petroleum. In these "fossil fuels," as in wood, is a wealth of energy from the sun, stored in the form of chemical bonds within the molecules, which can be released in the single step of ordinary combustion.

In most parts of the civilized world today, economists and ecologists are working

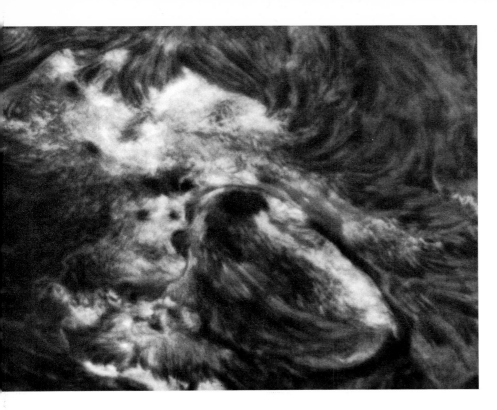

with agriculturalists in trying to find ways to get more food, fibers, and fuel for a rapidly increasing human population. They look to the green plants, which produce about 181 billion tons of dry organic matter each year beyond the needs in plant growth and reproduction. These 181 billion tons are available as nourishment for all other living things. Of the total, approximately a third comes from plants in oceanic and freshwater areas and two-thirds from plants on land.

Man has already turned toward his own use about one per cent of the current productivity from green plants. For any one species, out of at least a million kinds of living things, to control so disproportionate a share of the world's renewable resources is a phenomenon of the century. Yet each year at least 20 million people die of malnutrition. Since all of the lands that are suitable for raising food are already in production, the greatest hopes for a gain in nourishment for man seems to lie in increasing the efficiency of photosynthesis.

Under controlled laboratory conditions, plant physiologists have been able to get certain plants to incorporate into organic compounds about a fifth of the energy that reaches them. But this rate, which represents an efficiency of 20 per cent, is reached for only a few minutes or hours at the peak of the growth process. With algae, the best productivity that has been sustained for several days is an increase in weight amounting to 1 ton of organic material (dry weight) per acre per day. This represents storage of about 3.5 million kilocalories per acre per day and an efficiency of close to 10 per cent. It is feasible only when plants are given ample water with the right amounts of dissolved materials, an optimum temperature, continuous illumination at a tenth the intensity of full sunlight, an atmosphere containing 3 per cent carbon dioxide, and are evenly spaced so that none is shaded, and no sunlight passes unused.

Unfortunately, the efficiency measured under field conditions is nearer 2 tons per acre per year—less than 0.03 per cent efficiency. Silage crops may be efficient enough to yield 2.5 tons to the acre annually. A maize crop of 35 bushels to the acre, which represents about 2 tons of dry organic matter per acre, shows half its yield in cobs, leaves, stalks and roots. Both algae and aspen forests in Wisconsin produce at this rate. A single ton of hay or wheat plants (a fraction of which is grain) per acre is considered good. In most agricultural areas, efficiency is held down to barely 0.02 per cent because the growing

The self-nourishing green plants of the oceans sustain both themselves and all the consumers and decomposers of the marine world as well. In the Indian Ocean, the minute blue-green alga Trichodesmium heldebrandtii *(below)* sometimes becomes so abundant in aquatic windrows formed by prevailing winds that it is clearly visible from an altitude of 5,000 feet.

The glassy skeletons of single-celled radiolarians *(right)* are the remains of plant-eating consumers in the marine food web. Minute protozoans, they ingest the algae and other microscopic foods. The rubbery kelps *(center)*, such as those often washed up on Kerguelen Island, are among the coarsest of algae and an important source of oxygen, which they release through photosynthesis.

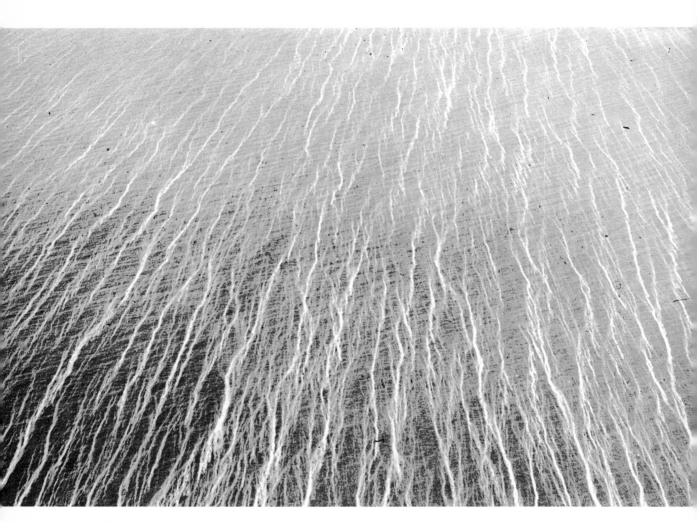

season averages only 100 days per year. Although the global average for all land vegetation is about 0.27 per cent, that for plant foods usable by man and domestic livestock is 0.0001 per cent.

The limitations in the rate producers capture energy has become evident in some measurements made near Adelaide, Australia, by Professor J. N. Black. He found that a clover field given the best seed and fertilizer yielded about 3.6 tons to the acre, instead of 29 tons, as calculated from the records of available sunlight. From mid-November to early April, the clover could not use the sunlight due to the summer drought, and the loss in efficiency amounted to 17 out of the potential 29 tons. From April until June, the plants had too few leaves to capture all of the light and thereby failed to produce about 1.3 tons. From late June until the return of summer drought, the leaves shaded one another, diminishing the total rate of growth by another 7.1 tons below the calculated potential. Much of the

achieved by sowing selected clover seed earlier in the season and at high densities in order to get leaves spread sooner and production to its peak rate. Only a combination of all these changes would raise toward a practical maximum the efficiency of photosynthesis in the field.

Nutrition of Plants

Not until the nineteenth century did scientists begin to study plant nutrition and the efficiency of crops. Among the first was the German chemist Baron Justus von Liebig. From his investigations into the relationship between crops and their soils came a fundamental principle of ecology known as Liebig's "law of the minimum." It states that the "growth of a plant is dependent on the amount of foodstuff that is presented to it in minimum quantity." Although water and carbon dioxide are required in large amounts by green plants, neither was included originally among the nutrients that von Liebig investigated. He tested, instead, substances such as calcium which are needed only in minute amounts but are often scarce in soil. He wrote in 1840, "The aptitude of the soil to produce one, but not another kind of plant, is due to the presence of a base which the former requires, and the absence of that indispensable for the development of the latter."

Von Liebig's classic *Chemistry in Its Applications to Agriculture and Physiology* includes a complaint: "Experiments have not been sufficiently multiplied so as to enable us to point out in what plants potash or soda may be replaced by lime or magnesia; we are only warranted in affirming that such substitutions are in many cases common." Now that the necessary investigations have been made, his generalization has found some support. The avail-

opacity of the leaves, moreover, was due to cells that gave support and provided conduction without directly contributing to photosynthesis.

The limiting effect of drought could be lessened through irrigation, but would be feasible only if the cost could be recovered through increased production. The total yield might be improved by cutting the leaves as soon as they began to shade one another and utilizing them to feed livestock. Far more improvement could be ability of one substance may modify the utilization, and hence the need, of another. Called "factor interaction," this supplementary principle applies also to the relationships between chemical and physical factors. Thus many plants (and animals) in estuarine waters can tolerate low salinity so long as the temperature is low, but will die if the water is warmed a few degrees. At a higher salinity they suffer no stress from the same increase in temperature.

Carbon dioxide is second only to light as the principal limiting factor in photosynthesis. Release of carbon dioxide from organic matter by respiration, decomposition, natural combustion and gas from volcanoes cannot match the demand. Manmade fires and deliberate combustion of fossil fuels during the last century have raised the concentration of carbon dioxide in the atmosphere less than might be expected, because the gas dissolves in sea water and there combines with calcium-ions to form deposits of limestone (calcium carbonate). It is estimated that 18,000 to 72,000 billion tons of carbon dioxide have already been withdrawn from circulation. By comparison, between 750 and 2,400 billion tons of gaseous carbon dioxide remain. This includes large concentrations in the air between the particles of soil. Another 4,500 billion tons might be released by complete oxidation of all remaining fossil fuels. Hence, no way is seen to improve the photosynthetic productivity of field crops and forests by supplying them with more carbon dioxide.

In some parts of the world, the soils are deficient in dissolved mineral substances. In the Philippines and other volcanic islands, for example, the extruded rocks that are the parent materials of local soils may contain almost no silicon. Just a few

handfuls of ordinary quartz sand, which is silicon dioxide, can change spectacularly the kinds of vegetation that will grow and the rates at which crop plants will produce. Plant physiologists sometimes demonstrate the role of this "micronutrient" by raising bean seedlings in plastic containers that have been rinsed and then filled with a silicon-free nutrient solution. This is prepared with all of the usual chemical substances dissolved in triple-distilled water, the last distillation of which was in platinum vessels. A crippling deficiency develops, the symptoms of which can be corrected merely by transferring the bean plants and their nutrient solution to a glass container. From it enough silicon dissolves to meet the normal needs of the plants.

To a large extent, the nutrition of the green plant determines not only how well it will grow but also its food value to other living things. The animal that eats a plant gets both a share of the energy the plant captured from the sun and also the plant's chemical materials that are necessary for its own growth and reproduction.

3 Consumers

Unlike green plants, animals and nongreen plants are dependent upon the presence of other kinds of life. They are not self-nourished but are "other-nourished" and hence called heterotrophs. They get their solar energy second-hand, use a small amount for chemical processes within the body and release the remainder as heat. Since herbivores go directly to the producers for their food, we consider them to be consumers of the first order.

Consumers of the First Order: Herbivores

Herbivorous animals generally rely upon particular kinds of food, for which they have a suitable digestive system that secretes the right enzymes for simplifying the organic compounds and releasing the chemical binding energy that came originally from the sun. Ordinarily, the herbivore stays and reproduces where acceptable foods are abundant. Yet if future generations of the herbivores are to eat, a breeding stock of the green plants must always be left reasonably intact. Ecologists are keenly interested in the many ways in which the plant-eaters avoid excessively exploiting their food plants. This branch of study is population ecology and the central problem is the regulation of the numbers of the animals.

The food of each herbivore must provide it not only with energy but also specific organic compounds. These include certain amino acids, vitamins, and a number of mineral nutrients. A few of the amino acids and vitamins are critically important because the animal has no way to store them. To a limited extent, an excess of one essential amino acid may reduce the quantity needed of another, but for synthesis of animal proteins to proceed toward growth, repair, and reproduction, all of the necessary amino acids must be present simultaneously. If one is missing for a moment or a month, the production of the protein stops until it is supplied. Even from one meal to the next, an imbalance in diet can cause a measurable change in metabolic rate.

A diet that contains the essential amino acids, the vitamins, calcium, and iron, ordinarily includes quite adequate amounts of potassium, magnesium, and phosphorus. For the latter three the needs of a green plant are as great or greater than those of an animal. In its growth, a plant absorbs from the soil the necessary potassium, magnesium, and phosphorus for itself and a herbivore. Since green plants have little use for sodium and incorporate only traces of it in their structure, herbivores may travel far to reach salt licks, where the soil or exposed rocks offer sodium salts.

In a few parts of the world, where soil and water are deficient in copper, iodine, manganese, zinc, molybdenum, cobalt, or selenium, herbivores show abnormal growth due to unsatisfied demands. In places where these elements are excessively abundant the plants may cause poisoning. Curiously, the nutrition of various plants and animals is so diverse that while one species can tolerate extremely low or high concentrations of a mineral nutrient, another finds the same situation quite intolerable.

Consumers of the Second Order: Carnivores

One check on the multiplication of herbivores is provided by the animals that nourish themselves at the expense of other animals. The spectacular predators are the game fishes, such as trout and bass, the constrictor snakes that suffocate and engulf their victims, and the birds and beasts of prey. Parasites, such as tapeworms, are far more subtle. Surreptitiously, they gain entrance to a host animal and generally live out of sight. Carnivores are consumers of the second order, because they get their share of solar energy third hand. Generally, they are highly adapted to searching afar for food and expend much energy in catching and killing it, or they excel, as most parasites do, in dispersing their young to places where food will be available. Between the carnivores and the vital processes of photosynthesis are two populations that can and do fluctuate—the green plants and the herbivores. Any disaster affecting the green plants decreases the number of herbivores the carnivores can find.

Predators and their prey and parasites and their hosts tend to live in uneasy balance. If the carnivores severely diminish the abundance of their potential victims, they themselves, or their offspring, starve. Death gradually reduces the population of carnivores, letting the herbivores multiply with less interference. Surviving carnivores find plenty to eat, and their reproduction slowly increases the population. Rarely does a carnivore destroy its food resource completely.

Most predators are far more versatile in their choice of food than was once believed. Owls eat longhorned grasshoppers and earthworms as well as mice. Foxes sustain themselves on plant foods when they cannot find enough small animals to eat. Ecologists regard these alternatives in diet as buffers, and see the advantage in having many such buffers available to tide the predators over periods when favorite prey is scarce. Many parasites are similarly safeguarded by infesting more than one species of host, a practice which lets a part of the population survive and reproduce when calamity strikes any one major source of nourishment.

Vultures are also carnivores, but have little effect on the population of the animals that comprise their food. They are scavengers and accept a wide variety of meats into their diet. In addition, they are valuable because they begin the process of freeing the chemical elements in the dead animals so they will be available for reuse in living systems. While policing the earth for usable remains, scavengers get their share of the solar energy.

Recently Lawrence B. Slobodkin of the State University of New York at Stony Brook expressed his "intuitive feeling that there are wise and foolish ways to be a hunter." He distinguished between a resource manager, who seeks a perpetual yield, and an exterminator, with no future interest in the prey species. His search for a "general theory of prudent predation" led him to recognize that two antagonistic operations must be neatly balanced through natural selection: a maximum capture of prey, and a maximum survival of the prey species.

In his laboratory aquarium, Dr. Slobodkin provided much small green plant food for herbivorous crustaceans, known as water fleas (*Daphnia*), and put in some carnivorous hydras (coelenterates) to see how they would fare. He sought to learn what rate of transfer of energy from the prey species to the predator would be the most efficient for long-term yield. If the prey were taken at a high rate, the survivors might increase their reproductive output to their full capacity because each would have plenty of plants to eat. But in this situation, the prey would become scarce; the preda-

Following the Energy Around

*Consumers and decomposers alike sustain
themselves on the solar energy bound into
organic molecules by the green plants
through photosynthesis. They also promote
the cyclic availability of raw materials of
which the world has a vast but limited
supply. The leaves of the deciduous trees
(center) capture solar energy and provide an
immense surplus of food that nourishes the
myriads of insects, such as the caterpillar
(top right) feeding upon a leaf. Grasshoppers
also get their energy second-hand by eating
blades of grass. As carnivores, spiders, vipers,
and owls get their energy and chemical
constituents from the herbivores. They, in
turn, die and decay, and, along with fallen
leaves, return their organic matter to the soil,
providing food for earthworms and all other
life in the soil. The processes of decay include
the decomposition work of bacteria.
Polypore fungi (above) speed the decay of
damp wood, particularly that of pines and
spruces.*

The decomposers prevent any great accumulation of dead organic matter. Fungi invade a dead tree from the outside (below), digesting through the bark and the most recent sheaths of wood before reaching the heartwood. The mold Aspergillus, which also serves in decomposition (right), forms colorless nutritive strands and masses of spores for dispersal by wind. Sometimes scavengers take a share of the energy and nourishment before the decomposers can do their work. Scarab beetles roll balls of camel dung (far right) to a suitable place for burial, where the organic matter in the dung provides food for the beetle larvae.

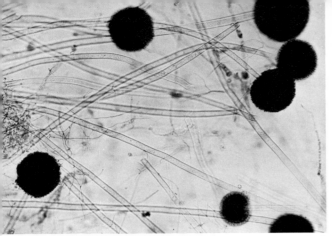

tors would have to engage in extended hunting, and relatively few of them could survive. On the other hand, if the predators took fewer and let the prey population increase until it was limited by the supply of green plants, the plants would suffer from exploitation by the prey and the total productivity of the food web be reduced.

By surveying the whole aquarium and following the relative changes in the numbers of water fleas and hydras, Dr. Slobodkin could decide what was the prudent course and adjust the rate of predation accordingly. He learned that the hydras could take between 8 and 12 per cent of the water flea population when the system was operating at peak efficiency. He sought to determine how an average rate of predation could work out with comparable advantages where, under field conditions, no individual predator could learn the state of the hunted population and behave accordingly. He realized that much of the answer lies in the age at which the prey species is killed. A predator can satisfy its need for food without unduly reducing the population of its prey if it takes individuals that soon would die. A young victim, for which the environment offers neither space nor food throughout its lifetime, is expendable. So is a sick individual. An old animal, past reproductive prime, can be eliminated with benefit to the prey species since the food and space the old one would use otherwise can go toward the needs of the young and reproductive. These expendable individuals comprise the categories upon which most wild predators actually concentrate their attention.

Consumers of Any Order: Decomposers

In a dead plant or a dead animal, or in dis-

carded leaves and feces, there are organic compounds containing energy available to many small organisms. Classified as bacteria, yeasts, and other fungi, these microbes that associate themselves with dead bodies are often referred to as the nongreen plants. They are distinguished from scavengers by their lack of a digestive cavity within which food is simplified for assimilation. Ecologists call them saprophytes, from the Greek words for rotting plants. They secrete digestive juices on the surface of their food and absorb the products of digestion.

The small size and relatively simple structure of the decomposers have both advantages and limitations. These organisms are well adapted for remaining dormant for weeks or years in order to conserve their meager store of energy until food is available. Many of them are carried from place to place by wind and water currents and are found almost everywhere. When conditions are right they can reproduce at a prodigious rate, sometimes doubling their numbers once or twice an hour.

Most decomposers are specialists with limited versatility. They generally operate in series, as a disassembly line. Wild yeasts, for example, fall as dust particles on fruit, become active and attack certain sugars. If oxygen is plentiful, the yeasts can release all of the energy in the sugar molecules with 12 or 6 carbon atoms and use most of it for their own growth and reproduction. Additional energy is lost as heat and the only chemical wastes are carbon dioxide and water. If oxygen is scarce, the yeasts work less efficiently. They still simplify the sugars and release both heat and carbon dioxide, but their wastes include ethyl alcohol, which contains substantial amounts of energy. The yeasts may continue the decomposition, or fermentation, activities until the food supply is exhausted, or until the concentration of alcohol rises beyond tolerance and the microbes are poisoned by their own waste.

Under natural conditions, the wastes from one decomposer rarely accumulate because they are used up as food by other decomposers. Vinegar bacteria, for example, get their energy by converting the alcohol from fermentation into acetic acid. Both of these organic compounds have two carbon atoms per molecule, but the acid has less binding energy. Different bacteria can get energy from the acetic acid, completing the degradation of the sugar in the fruit to useful energy, waste heat, water and carbon dioxide. Some ecologists regard as distinct those bacteria, yeasts, and fungi that gain the final residue of useful energy because they produce only more of their own kinds and inorganic wastes. They call these microbes transformers.

Virtually every organic compound that living things can synthesize is broken down by some chain of decomposers. A few compounds, such as the cellulose and lignin of wood and the chitin and keratin of animal bodies, are long-lasting because few microbes and animals have the enzymes necessary to digest them. Several organic man-made compounds have already become problems because they resist biodegradation and therefore accumulate. Several of the early types of detergents were in this category. They created foam problems in freshwater systems and displayed the extent of pollution. Modern detergents are all biodegradable, but the products of their decomposition vitiate fresh waters. Chemists see no possibility of returning to use of soaps, which the detergents largely displaced, because the market can no longer supply enough cheap fats for making the soaps that would be needed, and soaps do not remove dirt as efficiently as detergents.

Recently the chlorinated hydrocarbons, such as DDT and its derivatives, have become widely distributed. They offer advantages as insecticides because of their toxicity to pest species, their low cost, and their inertness to decomposers. However, they remain poisonous for years in "residual" sprays and dusts. Substitutes for DDT, costing more and endangering more kinds of life but lasting a shorter time because decomposers can destroy them, are being used increasingly.

All decomposers need time to dispose of

organic materials, even those from dead leaves, fallen branches, tree trunks, and other plant parts, the bodies of animals, and their feces. These sources of nourishment that the microbes can use are rarely as rich in energy as the living bodies of the green plants, the herbivores, or the carnivores. Correspondingly, the amount of energy in material to be decomposed is greater than the energy stored within the cells of live decomposers.

The importance of the rate at which decomposers can work is now evident in the oceans. Those that attack crude petroleum might be able to cope with twice the amount that leaks into sea water by natural processes. The current rate of addition may be five times as great. As a consequence of shipping about 1 billion tons annually of crude petroleum, and various products from refineries, about 5 million tons escape between continents. The drifting oil spreads over the interconnected seas, becoming an international problem. Vacationers complain about the fraction that is cast up on beaches. Commercial fishermen often find that the fish they catch are unsaleable because of oil. Peruvians and others see the cormorants and coastal waterfowl dying, their feathers coated with oil. For a decade or more the quantity of petroleum awaiting decomposition has been increasing. The distinguished French oceanographer Jacques-Yves Cousteau blames this and other pollution for an estimated decrease of 40 per cent in marine life during the last 20 years. Explorer Thor Heyerdahl, who grew enthusiastic over the purity of the ocean around his balsa raft *Kon-Tiki* as he drifted across the Pacific in 1947, reported that an almost continuous oily scum with masses of debris floated around his reed raft *Ra* on his trip from Africa to the West Indies in 1970.

The decomposers themselves are in grave danger, both in the soils and in the seas. They succumb to herbicides that are spread to kill weeds among crops or to maintain areas free of all woody growth. Ecologists fear the lasting consequences of these poisons. They shudder at the possibility (and statistical probability) that a ship loaded with herbicides will be wrecked in mid-ocean, freeing the toxic material that will kill most of the green plants upon which marine life depends. Too little is known of the consequences that could follow in a few months or years, depending on the rates of exchanges that take place among the ocean waters.

Still less has been learned about the long-term effects of chemical treatments that are being applied to land on an ever-increasing scale. Many (perhaps most) kinds of life are challenged by the powerful agents with which man is drenching the world. Generally they are applied for the sake of "controlling" (a euphemism for exterminating) vigorous vegetation and animal life of a few kinds. This pollution of the environment scarcely showed until the decomposers were overloaded or bypassed. It is considered the terrible price to be paid for trying to raise food and fibers with the efficiency of a factory.

4 Cycles, pyramids, and niches

In the few centuries since man learned to recognize chemical substances in the world around him, he has become aware that each element or compound that enters a living system is later released again. He can analyze the interactions among the producers, the consumers, and the decomposers by following the chemical pathways that link them all together. Each element actually follows a cyclic route, in which the nonliving world serves as a vast reservoir of raw materials. For this reason, the cycles are biogeochemical rather than merely biological.

The most obvious cycle is that of water, which is so often the limiting factor for life on land. Fully a quarter of all the solar energy that penetrates our atmosphere is used to evaporate water, chiefly from the oceans that cover about 71 per cent of the planet's surface. This amount of energy is at least 10,000 times as much as man now manages altogether. It converts to vapor about 210 cubic miles of moisture from the seas per day.

The life on land gains nothing from most of this evaporation, for the rainfall on the oceans returns about 186 cubic miles of water each day. But winds do propel water vapor equivalent to the remaining 24 cubic miles of water over the continents and islands, where about 62.5 cubic miles of dew, rain, and snow falls each day. About 38.5 cubic miles of water from the amount precipitated evaporates again. The remaining moisture is captured by plants, or sinks through the soil into underground streams, or flows through lakes and rivers, and supplies the needs of terrestrial microbes, plants, and animals. Daily, the moisture returns to the oceans from the rivers, restoring to them about 24 cubic miles of water, and completes the cycle.

The less obvious cycles have also been traced involving every element. But scientists pay special attention to those of carbon, oxygen, and nitrogen as volatile materials, and phosphorus and iron as nonvolatile. These five have had the greatest impact on human affairs and offer the key to understanding how nonhuman life replenishes the resources upon which all life depends.

The Carbon Cycle

Every organic compound contains both carbon and hydrogen. But the central event in getting these elements into combination is the capture (or fixation) of carbon from carbon dioxide through the processes of photosynthesis. Within living systems the carbon is recombined in many ways before it is finally released again in carbon dioxide through respiration, fermentation, or combustion. The withdrawal through fixation causes no measurable change in the concentration of atmospheric carbon dioxide because release of this gas occurs at an almost identical rate.

The annual rate at which aquatic plants fix carbon into organic compounds has been estimated by Dr. Bert Bolin of the University of Stockholm to be about 110 billion tons. The corresponding withdrawal by land plants would amount to another 38.7 billion tons of carbon. Release into the atmosphere in the form of carbon dioxide approximates 55 billion tons of carbon by decomposers in water, 50 billion by respiration of aquatic animals, 5 billion by aquatic plants, 27.5 billion by soil organisms, 11 billion by land plants and animals, and the remainder by wild fires, industrial combustion and volcanic activity.

Ecologists are agreed on the need to measure carefully the concentration of

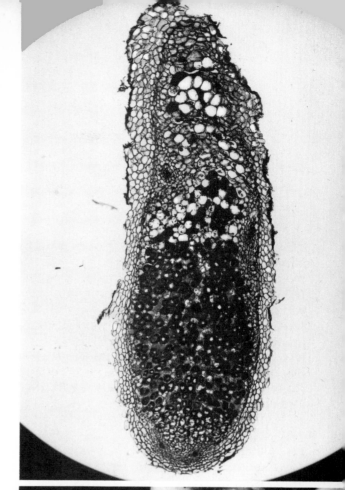

Microbes that are able to use dissolved nitrogen gas as a raw material in producing nitrogenous compounds are extremely valuable because they make available the nitrogenous nutrients needed by green plants. Those of the bacterial genus Rhizobium *(top), shown in cross section, live within the cells of a clover root nodule (bottom).*

atmospheric carbon dioxide and to avoid any change from its seemingly stabilized percentage. Yet opinion is divided on whether an increase in concentration would lead to a rise or a fall in the average temperature of the planet. Some ecologists fear that an increase in carbon dioxide will enhance the greenhouse effect and trap more heat, raising the temperature three or four degrees in a decade, causing the glaciers on Antarctica and Greenland to melt. Geologists calculate that this would raise the sea level all over the world by 200 to 300 feet, inundating many important cities and flooding all coastal plains. Another group of scientists agree that the change could come quickly, but predict that the warmer oceans would evaporate more water than at present, increasing the cloud cover and causing more snow to fall on polar regions, starting off another Ice Age.

No one has discovered yet how high the concentration of carbon dioxide was in the atmosphere during early Carboniferous times, before fossilization took so much carbon out of circulation. The luxurious growth of swamp vegetation may have been stimulated by a greater abundance of carbon for photosynthesis. Under experimental conditions, green plants grow faster and larger when the concentration of carbon dioxide is raised to as much as 5 per per cent. Higher proportions prove toxic.

In mixed deciduous and coniferous forests, carbon dioxide emanating from decomposers in the soil can benefit young spruce seedlings until they are about 15 inches tall. The gas comes up under the spreading branches of the low conical tree, increasing the efficiency of photosynthesis and helping the spruce grow despite dense shade. But as soon as the tree is higher than

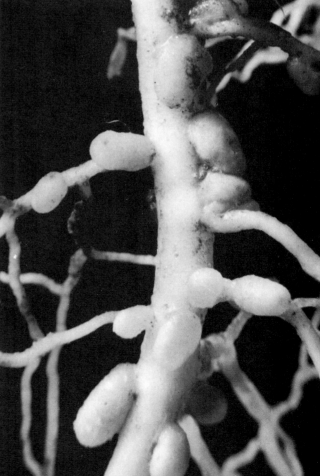

15 inches, air currents near the ground carry away the carbon dioxide too rapidly to sustain growth of the spruce. Limited by shade, it may grow scarcely higher for 30 to 40 years, then die. During this time, it can resume rapid growth if some adjacent tall tree topples and lets in light to the forest floor. The spruce then becomes a lone representative of its species in the midst of the forest, dispersing seeds that can colonize similarly.

The Oxygen Cycle

Oxygen follows a vital cycle that is less familiar to many people. In all aerobic respiration it serves as the acceptor of hydrogen, forming water. It is released again as gas only when the water is split by photosynthetic action.

The atmosphere currently contains about 1,160 trillion tons of oxygen. This vast quantity is apparently not only in equilibrium due to the respiration-photosynthesis cycle, but also the physical cycle that maintains ozone (O_3) in the upper atmosphere, and with chemical oxidation processes of many kinds. One of these converts to carbon dioxide the carbon monoxide given off by volcanoes and freed by incomplete combustion in man-made equipment. Another process is oxidative weathering, which produces compounds such as iron sesquioxide (Fe_2O_3).

Some years ago, William W. Rubey, now at the University of California in Los Angeles, made a rough estimate of the total amount of oxygen in the world. His figure of 65,000 trillion tons included 4,410 trillion circulating between the atmosphere (the 1,160 trillion tons), living things (an almost negligible quantity), and the sulfates and nitrates of sea water (about 3,120 trillion tons), while the great bulk (about 60,600 trillion tons) was "fossilized" in various forms. Carbonate rocks hold more than 54,000 trillion tons, sedimentary organic matter and sedimentary sulfates about 2,600 trillion tons each, and iron compounds (such as the sesquioxide, Fe_2O_3) most of the rest.

Dr. Rubey compared this total with the carbon in carbonates, sedimentary rocks and fossilized materials, in all living things, the hydrosphere and the atmosphere. The proportion of total carbon to total oxygen came out extremely close to 32 to 12, which matches the molecular proportions in carbon dioxide. More recent discoveries of iron ores in Australia and elsewhere, and better estimates of the oxygen content of sedimentary rocks, would make this proportion even closer. This supports the view that the oxygen in the atmosphere, the biosphere, the hydrosphere, and the sedimentary rocks was derived from carbon dioxide through photosynthesis. Today the average round trip of oxygen in the oxygen cycle through the atmosphere is believed to take about 2,000 years, as compared to about 300 years for the carbon in carbon dioxide.

The Nitrogen Cycle

The inertness of gaseous nitrogen makes it a particularly elusive nutrient for living things. It scarcely dissolves in water or in blood. And the little that does enter into solution is unavailable directly to all but a few kinds of life, despite their great need for it in organic compounds. Nitrogen is an essential component of every protein and every nucleotide. No microbe nor any living cell of a plant or of an animal can get along without these compounds. Yet until man learned to synthesize nitrates industrially, the whole world of life depended for its nitrogenous materials upon the very few kinds of organisms that can capture nitrogen and combine it into useful compounds.

During electrical storms, the action of lightning causes a small amount of nitrogen to join with hydrogen in forming ammonia (NH_3). The ammonia dissolves in raindrops, and some of it reaches the soil. Unfortunately, the amount that is made available to the roots of plants in this way is counterbalanced by the combustion of nitrogenous compounds and loss from the soil of gaseous nitrogen by the action of fires. Otherwise the cyclic movement of nitrogen from the atmosphere into living things and back again is the work of simple plants, specifically certain blue-green algae, bacteria, and soil fungi.

Some of these special forms of life carry the process all the way, absorbing nitrogen molecules and synthesizing nitrates or other nitrogenous compounds that higher plants can use as nutrients. These are the nitrogen-fixers, and their products are "fixed nitrogen." In wetlands and wetland agriculture, the blue-green algae are especially important; they contribute to the nutrients needed by rice plants grown in paddies. In grassland soil, the chief nitrogen-fixers are bacteria of the family Azotobacteriaceae. Better known as nitrogen-fixers are the bacteria of the genus *Rhizobium* in nodules on roots of plants belonging to the clover family (Leguminosae), and a fungus (*Streptomyces*) in nodules on the roots of various alders belonging to the birch family.

Decay bacteria in soil play important roles in the nitrogen cycle. Those that simplify amino acids and release ammonia, as *Bacillus mycoides* does, are said to carry on ammonification. They benefit some vascular plants, including rice that is grown in damp soil, because these particular kinds can use dilute solutions of ammonium compounds in lieu of other nitrogenous nutrients. Most other vascular plants need nitrates and get them through the activities of soil bacteria of several kinds that work in relays, carrying on effective programs called nitrification. Soil bacteria of the genera *Nitrosomonas* and *Nitrococcus* absorb either gaseous nitrogen or ammonia and release nitrites. Different bacteria, chiefly *Nitrobacter*, transform the nitrites into nitrates. Even within a single genus, the different species follow different courses in their relationship to nitrogenous compounds. The spore-bearing, rod-shaped, anaerobic bacteria of the genus *Clostridium*, for example, include nitrogen-fixers and also denitrifiers, which release nitrogen gas after decomposing nitrates.

The cyclic path of nitrogen between more than 4,000 trillion tons in the atmosphere and the small amounts fixed in nitrogenous compounds involves animals. The herbivores obtain amino acids and nucleotides through the action of their digestive enzymes on plant tissues. Carnivores and scavengers get these same nitrogenous compounds from the animal bodies they devour. In urine and feces the nitrogenous wastes reach the soil, generally as ammonia or urea or uric acid. These substances are absorbed by the roots of plants directly or after modification by soil microbes. The decomposers release additional quantities of nitrogenous substances as they simplify the dead bodies of both plants and animals.

Nitrogenous fertilizers are currently being synthesized and spread at a pace far exceeding the rate at which natural ammonification, nitrification and nitrogen-

Chemical Cycles

Benefiting from generous rainfall along a misty slope in Oregon, a forest of Sitka spruce with a carpet of ferns (left) carries on photosynthesis during the day, absorbing carbon dioxide and synthesizing organic matter while releasing oxygen. At night the same green plants absorb oxygen in respiration and liberate carbon dioxide. When the solar energy captured by the green plants has been fully utilized or lost as heat, the carbon and the oxygen have been recycled to the atmosphere. Carbon dioxide and water in outbursts from volcanoes tend, over millions of years, to increase the availability of these materials in the biosphere. All of the water evaporated from lakes, rivers, and oceans is ultimately returned to these sources through rainfall and the drainage of the soil.

*The great biogeochemical cycles
involving elemental carbon, oxygen,
and nitrogen, like that by which fresh
water circulates, take their energy
from the sun. They transfer materials
through living things from and to
vast storage "pools" in the
atmosphere, the hydrosphere, and the
geosphere.*

fixation occur. Although the manufacturing process may use hydroelectric power, the distribution of the fertilizer to and on the farms and the production of machinery for these purposes take their energy from combustion of fossil fuels. By the time the fertilizer has been made, bagged, shipped, and spread, more energy has often been spent than is gained in edible parts of the crop. This economic trade-off of energy in one form for energy in another has an ecological consequence that is causing consternation: loss of nitrogenous fertilizer into drainage streams, rivers, lakes, and eventually the oceans is overloading the nitrogen cycle. The agriculturalist may make a fine profit despite the waste of fertilizer. But organisms in the soil and in aquatic ecosystems gain only for a brief time, then lose through excessive growth on an unnatural schedule.

Cycles of Nonvolatile Nutrients

The producers of the world, whether cultivated green plants or wild ones, require significant amounts of the ions of nitrate, phosphate, sulfate, and chloride, together with those of magnesium, potassium, calcium, and iron. Generally both nitrate and phosphate are in meager supply. Calcium-ion affects the solubility (and hence the availability) of other nutrients, especially phosphate and iron. The calcium-ion, magnesium-ion, sodium-ion, chloride, and sulfate are often so abundant that they constitute limiting factors at the opposite extreme. By interfering with osmosis, they test the tolerances of plants and animals in terms of the maximum rather than the minimum.

Phosphorus is the element that most often limits growth of plants on land and soluble forms of iron is the most important

limitation in many parts of the oceans. The phosphorus is an essential part of each nucleic acid and nucleotide, and hence of the carriers of heredity and of energy in living systems. Along with carbon, hydrogen, oxygen, and nitrogen, phosphorus holds a place among the five most necessary elements for life. Together these five account for more than 95 per cent of the weight of all living matter.

A corn crop yielding 60 bushels to the acre absorbs about a tenth of the phosphorus in the top six inches of soil, and incorporates it in parts of the plant that are harvested. To raise another crop with similar yield, a farmer must replace the phosphorus. He may get fertilizer containing crushed phosphate rock, which is available on continents and islands where geologic forces have raised it from the sea. Or the phosphorus can come in guano, produced by sea birds, which get the phosphorus in the fish they eat, just as the fishes get it from the microscopic green plants, and these absorb it from solution in the sea. The guano with phosphorus of marine origin returns to the land less than 3 per cent of the phosphorus that rivers carry annually into the sea and that is precipitated there. At present the loss from the lands amounts to nearly 3.5 million tons annually. Yet the only other natural force that restores the phosphorus for terrestrial life is geological, in the unpredictable uplift of phosphate rock from sea into air. Much of the amount previously exposed in this way has already been quarried, crushed and used for fertilizer. The small remainder and all other known phosphorus in soil and strata of continents and islands account for less than one tenth of one per cent of the mineral matter that can be reached economically. It may soon be necessary to mine

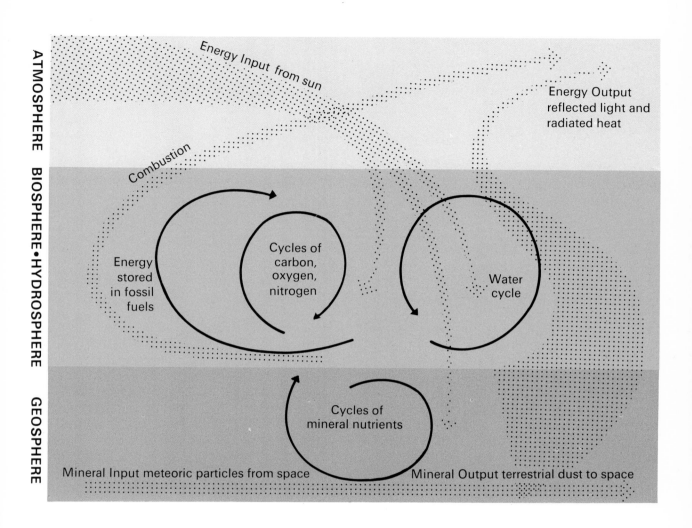

ATMOSPHERE BIOSPHERE·HYDROSPHERE GEOSPHERE

phosphate rock from the sea floor, just as guano is mined on offshore islands, because the supply of phosphate rock is running out. Ecologists and agriculturalists are beginning to realize that man's enterprises cannot wait for geological action, and that the supply of usable phosphate is the most critical limitation for the immediate future of terrestrial life.

In the oceans, the limiting factors that control the capture of solar energy are quite different from those on land. The producers tend to be suspended by turbulence at depths where their capture of light is relatively efficient. But because of the interface between air and water, the intensity of light varies daily on a schedule completely different for submerged plants from that for plants on land. Drought may never be a limiting factor, but low concentrations

of dissolved nutrients curtail production regularly.

Whenever they can, the producers in the oceans absorb ions containing atoms of carbon, nitrogen, phosphorus and iron in the proportions of 1,050 to 150 to 10 to 2. Generally, sea water contains about 10,000 atoms of carbon for every 150 of nitrogen and 10 of phosphorus, but only 1 of iron. Consequently iron tends to be limiting. Carbon, dissolved as carbon dioxide and in the form of carbonate and bicarbonate ions, becomes limiting only where upwelling or previous decomposition has greatly increased the availability of the other three. Under these special conditions, "blooms" of algae become noticeable and special significance can be seen in the respiration of microscopic cells at night and of decay bacteria.

About a thousandth of the total area of the oceans receives a free subsidy of natural fertilizer (including iron, phosphorus, and nitrogen) through upwelling. This area, which is less than 14,000 square miles, supports the guanay birds (cormorants) and the fishing industry of Peru, the penguins and surviving whales and the migrant arctic terns around Antarctica during the southern summer, the fisheries off the coast of South West Africa and in the Arabian Sea when the monsoon winds cause deep water to come to the surface. In these areas the average productivity of the green plants is about 13.5 tons per acre annually, and supports such a wealth of fishes that about half of man's total catch comes from these regions.

The rest of the commercial fishes are caught in the coastal zones, where a free subsidy of nutrients is added with reasonable regularity in the outflow of rivers and underground streams from the land. Although mineral malnutrition is frequent in these coastal parts of the ocean, the drifting plants there average annual productivity just under 7 tons to the acre. In less than 10 per cent of the total area, they contribute about 18 per cent of the product of the oceans.

The remaining 89 per cent of the marine world is open sea. Although just as well illuminated and exposed to the atmosphere, its productivity is limited to about $2\frac{1}{4}$ tons per acre annually by the scarcity of dissolved iron, phosphates, and nitrates. Most of the annual growth comes in early spring and ends when the nutrients have been incorporated into living cells. Thereafter the open sea becomes a biological desert. It waits month after month, not for a rain but for the decomposers to release the nutrients from the bodies of the dead.

Ecologists refer this to control over the growth of green plants as regulation by decomposition.

Pyramids

A different pattern among living things becomes evident when attention is focused on the energy relationships rather than on the nutrient materials that are cycled through life. The energy flows along channels limited by the nutritional characteristics of animals. It always enters an autotroph of some kind—a green plant or producer. The plant uses part of the energy it captures for its own respiration and growth. When the plant dies, the decomposers inherit the remains. By then, a part of the plant may have gone into a herbivore. As a consumer of the first order, the herbivore is on a separate level of ecological organization. The ecologist speaks of it as the second trophic level, the first being occupied by autotrophs. The herbivore is heterotrophic, as is a carnivore, yet intermediate in position. Carnivores occupy the third trophic level in this sequence.

On land, the amount of energy contained in the bodies of all producers together is ordinarily much greater than that in the bodies of all herbivores present. Correspondingly the energy of the herbivores exceeds that in all of the carnivores that prey upon or parasitize them. These energy relations form a pyramid, with a large base composed of the producers, a proportionately smaller tier to represent the herbivores, and a third tier of still less volume to indicate the carnivores.

Solar energy enters only the lowest tier of the pyramid of energy relations. Progressively the energy is degraded to radiant heat in a sequence that is characteristic of every chemical reaction carried on in life. The heat escapes from every trophic level and from the decomposers too. Ordinarily, the total radiation released to outer space each year equals the total of solar energy captured by the green plants. Any imbalance shows as a gain or a loss in the total energy in temporary storage within the organic compounds of living things and their still undecomposed remains.

The concept of the pyramid is often extended in relation to land life. The true producers in a forest are seen to be the individual leaves, just as on a prairie they are the grass blades that capture energy from the sun. The producers are enormously abundant by comparison with the number of herbivores. The populations of carnivores, including predators, parasites, and scavengers, are generally much smaller. The top predator, which only parasites attack, is a rather lonely individual with few near neighbors of its own kind.

A young American ecologist, Raymond L. Lindeman, attempted in 1942 to find the shape of the pyramid from quantitative relationships among the trophic levels. To stimulate research that might discover what fraction of the resources at each level was actually used by animals of the next higher level, he suggested some estimates based upon a prairie. Grazers might take between 10 and 20 per cent of the energy (or the numbers or the bulk) of their potential food plants. Carnivores might eat between 10 and 20 per cent of the grazers. Other ecologists accepted his challenge and sought to test what became known as the "10-percent law."

In a study of this kind at Silver Springs in Florida, Howard T. Odum found the principal producers to be beds of freshwater arrowhead (*Sagittaria*) with attached algae. They served as food for a number of different snails, aquatic insects, herbivorous fishes and turtles. Predatory invertebrate animals and fishes, including the young of large predators (chiefly bass and gar), were the principal carnivores. Bacteria and some fungi attended to the decomposition of dead plants, dead animals, and feces.

A similar study by John M. Teal had as its site a saltmarsh along the coast of Georgia, where the cordgrass (*Spartina*) grows a little even during winter, and the other producers (chiefly diatoms, flagellates, and other algae) thrive about equally in all seasons. Grasshoppers and many other insects eat the cordgrass; roundworms feed on its roots. Different animals feed on the algae and the particles into which the cordgrass breaks soon after it dies. The particles are detritus, partly the product and to a major extent the food of the bacteria and other decomposers. Bacteria and detritus are swallowed together by fiddler crabs, snails, and fishes such as mullet and menhaden that swim up the tidal gutters. Mussels and oysters filter out detritus particles that are suspended in the water washing to and fro.

These ecological situations can be compared quantitatively in terms of the solar energy available, the amount that is not absorbed, the energy that green plants capture, the energy they use for their own respiration, their gain in weight as shown by weight change and reproduction, the energy used by each trophic level of consumers, and the amount (if any) that remains as net productivity. For uniformity, all of these can be expressed in kilocalories of energy per square meter per year. (See chart on page 44.)

At Silver Springs, the submerged and emergent plants capture about 1.2 per cent of the light available, and store about

Productivity in ecosystems, expressed in kilocalories per square meter				
		Silver Springs		Georgia
Input of solar energy		1,735,000		600,000
Unabsorbed by photosynthesis		1,319,000		564,000
Gross primary production		20,810		36,380
Producer respiration		11,977		28,175
Net primary production		8,833		8,205
Total energy used by consumers		6,821		4,534
Herbivore respiration	1,890		596	
Carnivore respiration	331		48	
Decomposer respiration	4,600		3,890	
	6,821		4,534	
Output in new productivity		2,012		3,671

5 per cent of what they absorb. In the salt-marsh the vegetation gains from 24 per cent of the light input, but loses more energy in respiration. This decreases sharply their net productivity and thereby the food that is available to herbivores.

The consumers in the warm waters of Silver Springs use more of the energy they get for respiration and locomotion. In terms of the plant food that supports these organisms, the herbivores spend 21.4 per cent of it, the carnivores 3.8, the decomposers 52.0, making the total energy dissipated by the consumers total 77.2 per cent. In the saltmarsh, by contrast, the herbivores use up in respiration and locomotion only 7 per cent of the energy the green plants store; the carnivores use up 0.6 per cent, and the decomposers 47, making the total 55 per cent of the net production. The balance in each case, 22.8 per cent at Silver Springs and 45 per cent in the saltmarsh, is new organic matter that can be exported. It includes living emigrants who may become colonists elsewhere, and food for organisms at a distance.

The herbivores are more efficient in their way of life than was anticipated. At Silver Springs, from the 8,833 kilocalories per square meter of nourishment available to them in the green plants through the year, they captured 3,368 into their own gross productivity. This accounts for more than 38 per cent. In achieving so much for their own welfare, they must have devoured almost half of the vegetation available,

losing the residue as undigested and un-absorbed material from plant tissue whose death they caused. The energy in these residues, mostly as feces, nourished the decomposers.

If we think of the herbivorous fishes, the turtles, the numerous aquatic insects and the snails in Silver Springs acting in concert to eat almost half of the plant materials there, we realize how close they came to destroying the resource upon which the future of their kinds depends. The living remainder of the vegetation was needed to reproduce itself and to maintain the fundamental productivity.

No single factor in the environment regulated the activities of the herbivores, preventing them from overeating and destroying their resources. From Dr. Odum's measurements, we see only how much of the energy they captured was used up in their own respiration, and how much turned up in the gross productivity of the meat-eaters that ate the herbivores. The plant-eaters used up 1,890 of the kilocalories per square meter out of the 3,368 they gained from their diet. This is more than 56 per cent. It includes the respiratory requirements during hours spent doing nothing, staying quiet and inconspicuous while predators moved about close by. For the insects, it represents the fraction of the total life span spent in moulting the exo-skeleton and waiting for a newly exposed one to harden. It encompasses also the energy actually used by disease organisms inside the herbivores, reducing their vigor and their speed in traveling to new supplies of food.

The remaining 44 per cent of the energy the herbivores got from the plants showed up in organic compounds. But of this 37 grams per square meter (an ounce per square yard), the small carnivores took their toll. They retained from it energy amounting to 383 kilocalories per square meter after the nourishment had been absorbed into the bodies of the second-level consumers. How many kilocalories worth of herbivores lost their lives to supply those 383 kilocalories of absorbed food? If the carnivores were 50 per cent efficient in their killing, they would take 766 kilocalories from the 1,478 of net productivity per square meter among the herbivores. Barely 52 per cent of the organic matter in the herbivores would be left as live individuals able to sustain the population and carry on the heritage. Violence or the imminence of violence from predators reduces the impact of herbivores on the various plants.

The successful herbivores were those that found opportunity to eat enough for growth and reproduction and that escaped death from disease and predators long enough to do so. Although few in numbers, leaving the green plants a chance to maintain themselves and export a few offspring, these herbivores did not disappear. They kept the carnivores alive, tolerated their parasites, and had a little energy left over. The balance went into new lives to replace the old ones that came to an end, and a continuing output of potential colonists. Some day the operations of natural selection might offer the adventurers a new place in the world.

The large output of wastes from herbivores is understandable. These animals cannot be more efficient in using plant foods because the materials they eat contain a large bulk of indigestible cellulose or silicious matter, an oversupply of potassium compounds and scanty amounts of sodium. The herbivore must process far more plant food than it needs for energy if it is to accumulate enough of certain amino acids that are scarce in vegetable matter but essential for the synthesis of animal proteins.

Most carnivores encounter no comparable difficulties in converting their food into energy and components for growth. Their efficiencies are correspondingly greater. Yet the proportion of the energy they get that can be used for growth and reproduction is small because they must expend so much in getting the next meal or infecting another host. The scarcity of suitable prey or carrion is generally critical. An active predator, such as a weasel, may lose 93 per cent of the energy it captures in respiring and running around in hunting activities. Its net gain—a meager 7 per cent—may be enough to keep its kind in existence, but not to support another trophic level in the pyramid. A predator that lives on predators is an ecological luxury that few parts of the world can afford.

In the open ocean, predators do feed on predators. But there the whole concept of pyramids requires modification because the producers are mostly microbes. Except in a few places where the nutrition of the green plants receives an important subsidy through an upwelling of water rich in nutrients, the herbivores are mostly microscopic too. Carnivores come in a whole range of sizes, each nourishing the next larger, from copepod crustaceans smaller than a grain of rice to arrow-worms and fish larvae an inch long, to fishes and squids less than a foot in length, to tunas and voracious sharks.

Organisms in the small size range need to accumulate only a minimum of energy and of nutrient materials before they can reproduce. They pass on their energy and chemical components to descendants and to higher trophic levels without developing much bulk, accumulating impressive stores of energy, or showing spectacular populations. If all the producers in a vertical column of sea water were collected from a region of moderate productivity and compressed to a layer comparable in density to that of a green leaf, the layer would be scarcely as thick as a sheet of writing paper. In regions of high productivity, the total weight of the population and the abundance of individuals may show a gain of two or three times, but rarely more. When productivity is high, so is the turnover rate. The energy passes onto herbivores and carnivores, in which the lifespan is longer and the total weight does show a marked increase.

To the amount of energy that must be stored in organic compounds before the members of a species are ready to reproduce, Eugene P. Odum has given a useful name: the work loop. It is the fundamental feature that determines the average length of time, perhaps centuries in a tree, decades in an elephant, or days in a decomposer, between the first accumulation of organic substances in the growing individual and their final simplification by decomposers. Only through this turnover do other organisms get their energy and their nutrients. Over the years it is the work loop and the turnover rate that determine the productivity.

In subdividing the pattern of energy flow into trophic levels, oversimplification is inevitable. We are tempted to think of the microscopic drifting plants of the open ocean as producers that are ecologically equivalent to the grass blades on a prairie or the leaves on forest trees since each of these is a center of photosynthesis. In some places, the correspondence is convincing. Along the Peruvian coast, for example, the principal producers are colonial diatoms that cluster in masses as big as a pea. Anchovies and other fishes are the obvious herbivores, gulping down the masses of plants or filtering out the smaller clusters in the surface waters. The famous guanay birds and pelicans and sea lions are the carnivores. These levels of consumers can be compared easily to the grasshoppers and prairie dogs on a grassland, the caterpillars and boring insects of a forest, the hawks and coyotes on the one hand, and the insectivorous birds among the trees on the other.

The decomposers in the sea are primarily bacteria, and clearly equivalent to the bacteria and fungi that serve this role in soil. On land, however, the inconspicuous decomposers associate with soil-dwelling herbivores (such as collembolan insects and mites) and soil carnivores (such as other mites and nematode worms). Together these organisms, which often are ignored, tidy up all of the other trophic levels, recirculating the chemical elements and restoring the energy balance of the planet by freeing radiant heat. Including them and the scavengers is essential in any adequate account of ecological relationships.

When the transfer of energy along the myriad pathways among the plants and animals in a region is examined, measurements must be made piecemeal, then combined into a meaningful pattern. Generally the flexibility within the pattern resembles that in the electrical network of a modern electronic circuit, because what happens to energy flow in one part of the system affects events elsewhere. Howard T. Odum has actually tried to construct electrical models of ecological systems. As he varies the conversion of electrical energy into radiant heat in the linked parts of the network, he can read from his meters the changes that are analogous to nutritional adjustments in the real world of living things.

No model can imitate adequately the actual relationships among the consumers because animals of most kinds show considerable versatility in changing from predation to scavenging or, temporarily, to a diet of plant material. Not only may animals shift from one trophic level to another and be omnivorous in a single year, but also may progress through regular changes in type of food during their normal growth. The transformation of a herbivorous tadpole to a predatory frog is the converse of that from an insectivorous nestling of a sparrow into the seed-eating adult. The larval flea scavenges for animal fibers, often in the bedding of the dog on which the adult fleas probe parasitically for blood. The larval mud-dauber wasp has no choice but to feed on the spiders with which the nectar-sipping, pollen-eating adult insect has filled the private cells.

Niches in the Food Web

Ecologists have learned that under wild conditions the trophic relationships are rarely simple enough to be called a food chain, such as from a leafy shrub to the deer that browses on it, and then to the cougar that kills and eats the deer. The deer eats many kinds of plants, and chooses a different diet in each season of the year. Other herbivores eat these same plants and others that a deer ignores. The cougar may prefer deer, but preys also upon moose, elk, and domesticated horses if they are easier to catch than smaller animals. The living things in a region form a complex web of energy relationships, commonly known as a food web.

Within a food web, each species ordinarily relies upon a unique and adjustable selection from among the foods available. By being unique, it avoids direct and full competition with every other species among its neighbors. By being adjustable, it can rely more heavily on alternative resources when the supply of its favorites diminishes temporarily or permanently.

Each species or kind of living thing can be described according to its particular selection of foods and of places to live. These characteristics match the features that help it survive in its way of life. To this combination of energy resources and geographic site, the distinguished English ecologist Charles S. Elton gave the name ecological niche. Each niche is more than a subdivision of a trophic level. It includes also a share in the physical environment, such as the local topography and climate where members of the species live. Each niche overlaps with many others, which belong to neighboring species; but it coincides with none.

The niche of a North Atlantic lobster, for example, includes a place for hatchlings among the minute life drifting at the surface of the sea, where larval lobsters are herbivores while feeding on microscopic plants. Later the niche includes a place among the crannies on the sea floor over a continental shelf, where the transformed lobster—no

The Niches of Animal Life

For many animals, normal development requires a change of habitat and way of life according to the time of day or year or life cycle. The immature mayfly (top, far left) lives in a freshwater stream but as an adult has a brief opportunity to fly and mate in air. The guillemots (far left) of European coasts remain at sea, feeding on fish except at nesting season, when they fly to inaccessible cliffs to raise their young. Earthworms (top, left) must have loose, porous soil containing organic matter in which to live. They must emerge at night to find mates and exchange sperm cells for mutual fertilization. The kangaroo rat (left, second from top) could not survive in the desert without a burrow in which to escape the heat, and must emerge at night to find food. A red fox roams widely but needs a secure den in which to tend its young (left, third from top). The sea lion and marine iguana (left, bottom) on a Galápagos shore are able to bask together since they seek different foods—fish and kelp respectively—in the underwater world. A tubeworm lives as an adult in its spiral shelter on the sea floor more than 1,700 feet below the surface (above) but goes through its larval stages drifting among the plankton.

longer a larva—scavenges or preys upon animal foods of many kinds.

Habitats and Ecological Communities

Often the place where a plant or animal is characteristically found is called its habitat. This may be a geographical site that is recognizable from information about the nonliving environment. Or the habitat may be described in terms of living things such as a grassland, a marsh, or a forest edge where deer are found.

The associations that are sustained over long periods by living things impressed Professor Karl A. Möbius of Kiel University in Germany while he was studying oysters and oyster culture along the coast of the Baltic Sea. He noticed how regularly the sampling dredge brought up certain kinds of snails, worms, and echinoderms as well as oysters. He found the same species associated with oysters in the estuaries of English rivers. Recognizing that these species must share more than just a habitat, he proposed in 1877 to call this "community of living beings" a biocoenosis. Later his word was largely replaced by a simpler phrase, the "ecological community." Each community consists of many populations of separate species which interact in a web of food relationships.

Sometimes the energy pyramid within an ecological community can be constructed from writings and records that were kept for other purposes. In New England three centuries ago, observant men found a forest community in which the populations of trees seemed endless. White-tailed deer browsed on shrubbery along every forest edge, preyed upon by cougars, timber wolves, and small numbers of Indians. Bobcats, foxes, and coyotes killed a good

many fawns in season, but otherwise had little effect upon the deer population.

Each cougar tended to stay by itself or with its young in places where it could catch a deer every other night. The colonists came to expect one of these big cats on every major hill, and for this reason they called it a catamount—cat of the mountain. Often the small-headed, long-tailed animal measured nine feet in length and weighed 200 pounds. It defended against invasion by other cougars a hunting territory about nine miles across, patrolling its boundaries or crouching in wait for a passing deer. Generally the cougar leaped from a low limb or a big boulder upon its victim, then satisfied itself with seven or eight pounds of venison. Satiated, the cougar covered the remains of the carcass with leaves, actually abandoning it to the smaller predators that were always scavenging for meat.

Some of the deer the cougar missed were run down by the local pack of wolves. A family group of four wolves, perhaps a 150-pound adult male, his 80-pound mate, and two 75-pound youngsters, might live on the same area as one cougar. They could ignore the cougar's nocturnal forays, for they chased their prey by day.

The wolves would gorge themselves at a kill, eating as much as a fifth of their own weight at one meal. If the prey were large, such as a 150-pound deer, each wolf would dig a hole, disgorge its first meal and return for an equally generous refill. Wolves and other members of the dog family are used to eating disgorged food when hungry, for this is the way parents ordinarily provide meals for their weanling young in a den. Together the members of a pack would clean up a carcass, leaving little to attract or nourish other carnivores. One full-grown deer might satisfy the pack for a week, and they would feel no urge to kill another.

An area nine miles across could be home to one cougar, four wolves, various small predators, and about 768 deer if each deer had 50 acres to browse in. This number of big herbivores would cause no serious damage to the producers. But to prevent an increase in the deer population, about a third of them would have to be removed annually in one way or another. Under natural conditions, the cougar would take about 183 of the 259 expendable deer, the wolves 52, leaving about 24 to be accounted for by an assortment of bobcats, lynxes, foxes, coyotes, and occasional Indians. The food web, like the pyramid of energy relations and of numbers, would be in balance. It would provide for a good many smaller carnivores which ate venison when they could. Otherwise they depended on rabbits and hares, squirrels, mice, voles, such birds as they could catch, or as a last resort, plant foods of various kinds.

Generally each ecological community is named according to certain conspicuous plants or animals (or both) that appear there. The northern coniferous forest is a spruce–moose community, and many arid parts of the American Southwest a chaparral–jack rabbit community.

On land, ecological communities are distributed geographically in a pattern that reflects climatic features, particularly rainfall or other precipitation during the year and the normal range of temperatures. Heavy rainfall sustains a rain forest of spectacular density. It may be a tropical rain-forest community where the temperature never approaches the freezing point, or a temperate rain-forest community if frost is a frequent hazard. Where the rainfall is moderate but winter snowfall is heavy,

yet the summer is warm enough to thaw the soil completely, the Northern Hemisphere has forests of lesser density. They are deciduous where winter is of moderate length, but composed of evergreen conifers where winters are particularly long. A summer so short and winter so cold that only the top few inches of the soil thaw each year limits the region to tundra types of life, whether at high latitude or high altitude. A moderate rainfall or snow cover will support the growth of a thornscrub community if the summer is hot and dry. Less rainfall and a similar summer match the adaptations of many grasses and grass-eaters, and produces a prairie, steppe or savanna. Scanty rainfall and a hot dry summer can be tolerated only by desert flora and fauna. Each of these combinations of living things and climate is called a biome: the desert biome, the thornscrub, the grassland, the tundra, the northern coniferous forest (known by its Russian name as the taiga), the temperate deciduous or mixed forest, the temperate rain forest, and the tropical rain forest.

In lakes and oceans are aquatic communities that rarely suffer from drought or cold, being protected from quick changes in temperature by the thermal inertia of water. They show no broad subdivisions comparable to the distinct biomes on land. Instead, they are affected by physical and chemical factors peculiar to the watery environment. Along coasts the marine community may be strongly influenced by the tides, and battered at intervals by waves due to storms. Dissolved mineral substances vary in concentration over a wide range, from so low as to limit plant growth in alpine lakes, to the minute amounts characteristic of fresh waters at low elevations, and to the salty solutions found in

the sea. The amount of suspended matter affects the depth to which light can penetrate at intensities great enough to support photosynthesis. Standing waters may be lasting or temporary. Flowing rivers offer their own special hazards to living things, carrying any that are not attached or able to resist the current from fresh water into the sea. Rarely can a member of one aquatic community survive being transferred into another.

Close study often reveals that each species in a community occupies only small parts of the available habitat. The birds in a forest tend to be stratified: warblers that build and pursue insects in the tree tops meet quite different climate and food from the ovenbirds that nest and forage close to the ground. The subdivisions of a single habitat can be called microhabitats, each with its microclimate.

Ecosystems and the Ecosphere

Each ecological community and its non-living environment interact in a systematic way, and constitute the fundamental unit of ecological organization. It may be a temporary pond, a permanent lake, a forest tract, a patch of prairie, or of barren tundra. If it gets little energy or raw materials from surrounding areas, it can be conveniently examined and measured as an ecosystem.

That living systems could be essentially closed, and that each species in them could be understood adequately only by considering its relations to all others, was pointed out first in 1887 by Stephen A. Forbes, an economic entomologist in the employ of the Illinois Natural History Survey. He justified this conclusion in an article, "The Lake As a Microcosm." Today his word microcosm has been replaced by the term

ecosystem, proposed in 1935 by the British scientist Sir Arthur G. Tansley. The analysis of an ecosystem differs from that of its community in that the physical and chemical features of the environment are explored too, particularly the energy budget and interchanges of chemical substances.

The functioning of ecosystems causes physical and chemical changes in the non-living components of the earth, affecting the nature of the soil, the chemical composition of the atmosphere, and the climate. Yet each ecosystem is a local system, with a limited effect. When the cumulative effect of all of the world's ecosystems is considered at once, as in trying to measure long-term changes in the earth's atmosphere or the chemical constitution of ocean waters, the ecological concept is correspondingly broadened to become that of the ecosphere. Ecosystems are thus parts of the ecosphere.

Sometimes ecologists arrange their categories in an ascending sequence from smallest at the bottom to the largest at the top:

Ecosphere
Ecosystems
Communities, each with
many habitats, those on land
in a definite biome
Populations (species or subspecies,
each occupying its own ecological niche)
Individuals (organisms)

Each of these levels has its stabilizing features and its disruptive forces. The study of ecology is largely concerned with discovering how living things use solar energy to maintain the complex organization that has evolved.

5 The physical challenges to life

The world may be contemplated as though the physical features of the environment were independent of the chemical features. Yet, as in a tapestry whose design is a consequence of interwoven warp and woof, the physical and the biochemical world become significant only when the two are considered in combination. Both are modified, moreover, by the presence of life.

The physical world shows measurable differences in temperature, pressures, and currents in the medium, whether air or water, used by living things. Some of these differences continue for millennia. Others follow a rhythmic pattern according to the season of the year, the time of day, or the ebb and flow of tides. The extremes in physical conditions often influence evolution because they tend to select for survival those plants and animals that possess the greatest natural tolerance.

The Limiting Effects of Temperature

In Yellowstone National Park in the United States, and in Iceland and New Zealand, thermal springs support several kinds of plants and animals in water so hot that the human hand is scalded if immersed in it for just a few seconds. Where the water is a few degrees cooler, other kinds of life find a place.

Along the fringe of snowfields in Banff and Jasper National Parks in Canada and in similar localities in Japan, strange little black insects of the order Orthoptera take shelter around the edges of stones. When chilled, these insects of genus *Grylloblatta* become inactive; each year they withstand being frozen and thawed repeatedly. Whenever the temperature rises a few degrees above their freezing point, they creep about and scavenge for vegetable matter that decays at an extremely slow rate

because of the cold. Yet if one of these insects is held for a minute or more in a human palm, carefully shaded from the sun, it dies of the heat.

Living things in thermal springs and near snowfields tolerate extreme temperatures. Yet each can be active over a narrow range. The adjective stenothermal is applied to these forms of life because of their narrow temperature tolerance. Organisms with a broad range of tolerance are said to be eurythermal.

No matter what range in the environmental condition is tolerable to the organism, it develops signs of being under stress if it is subjected to either extreme. The American ecologist Victor E. Shelford, who investigated this phenomenon, called it the "law of tolerance." But changes with time, both in the environment and the organism, generally shift the point of stress from one critical factor (such as temperature) to another. For this reason, the Canadian ecologist Pierre Dansereau placed first among his twenty-seven "propositions" of ecology his "law of the inoptimum." It states that "no species encounters in any given habitat the optimum conditions for all of its functions."

Physiological stress has a time limit. A small stress for a long time may be equal in effect to a major stress for a short time. Sometimes, however, a small stress leads to adjustment within the individual. It may cause a chemical change that reduces the stress by appropriately shifting the optimum zone. A plant or animal that would be killed by chilling close to the freezing point in midsummer can often develop hardiness to cold before winter comes. Often this adjustment is matched by a loss of tolerance for hot weather.

Many animals protect their lives by mov-

ing from one place to another as soon as stress becomes severe. Behavioral ecologists (or ecologically-minded ethologists) notice that in arid areas of the American Southwest the horned lizards (*Phrynosoma*) limit their exposure to the direct sun after sunrise and before sunset according to the sun's heating effect. As soon as it warms their bodies to the point of physiological stress, they burrow underground and avoid a further rise in temperature. In early evening, they reappear but, until the air temperature decreases, they dash into the open only briefly to catch insects.

Then they begin tilting their flat bodies toward the sun, thereby capturing extra warmth and extending their active day. When this fails to prevent physiological stress toward the low end of their optimum range, they hide underground for the night.

On any given day, most individuals show a preference for conditions near the midpoint of the optimum range. A few are more tolerant of low temperatures in the normal range, and a few of high. Air-conditioning engineers find this variation among people. At a reasonable setting of the thermostat, 60 per cent may be satisfied while 20 per cent complain of being too cool and the other 20 per cent of excessive heat. Any decrease in temperature places those who prefer warmth under stress; any increase similarly affects those who prefer cool surroundings. Under wild conditions, plants and animals that are under stress tend to disappear. Mutants with a genetic constitution that reduces the stress are favored, leading to adaptations that are inherited. Whole species that fail to adapt in this evolutionary way can be expected to become extinct.

Beyond the tropics, the temperature tends to vary irregularly but cyclically over a considerable range between summer and winter. Both plants and animals may match these changes by reducing activity during

at least one season. Organisms that become dormant during the warmest part of the year are responding to drought more than to heat. So, in a way, are many that show winter dormancy since low temperature limits the availability of water to plants, and a reduction in plant growth limits the food for animals.

Irregularities in temperature require special adaptations too. Plants and animals must not respond to a warm week in autumn or early winter while they wait for the warm weather of spring. Warmth alone is not enough. The safety system consists of a requirement for a month or more of significant chill before the body can respond to a progressive rise in temperature. In plants this phenomenon is called vernalization, and in animals diapause. The buds on a lilac or the moth pupa in its cocoon may slip the safety catch if kept in a refrigerator through October and then exposed to artificial warmth in November. Under normal conditions in a wild population, only a few individuals react to shorter periods of chill and subsequent heat. Natural selection eliminates them from the breeding population, while retaining those that show the slower reaction.

A similar mechanism prevents in most years the germination of ripe, fallen seeds from shrubs such as the Californian lilac (*Ceanothus*). After a warm summer, the brief intense heat from a brush fire has no effect. Any seedling that started to grow then would most likely die of drought or cold before it grew large enough to gain resistance. But after a cold winter, similarly intense heat sensitizes the seeds to moisture. At the first rain they sprout, take root, and spread their green leaves. Before the summer drought arrives they may be well established, their stems sturdy, the bark

thick, the leaves able to tolerate considerable desiccation.

Fundamentally, these adaptive features are carried from one generation to the next in the genetic code borne by the chromosomes in each cell. They express themselves in chemical reactions at the cellular level, and in the schedule with which these actions are turned off and on by the inherited determinants. This is particularly evident among the familiar insects, which possess just a single stage in their life history when cold weather is tolerable. For a praying mantis or a tent caterpillar moth or an aphid, it is the egg that remains exposed all winter. For the tiger moth with the best-known caterpillar, it is the woolly bear larva, curled up somewhere among the fallen leaves. For the cecropia moth and the luna, it is the pupal stage, snug in a silken cocoon. For the mourning cloak butterfly, the bumblebee and the white-faced hornet, it is the adult which finds shelter as best it can beneath loose bark or low among the stones of a dry wall where deep snow provides natural insulation.

Each insect develops at a normal pace until it reaches this special stage in autumn. Then it stops, entering its specific diapause, and waits until a period of cold makes it receptive again to warmth.

The monarch butterfly (*Danaus*) and a few other kinds of insects are unusual in avoiding winter chill by flying toward the Equator at the same season that migratory birds and bats are winging in similar directions. These behavioral adaptations are timed by changes in the length of night, a phenomenon known as photoperiodism. They all avoid periods of food scarcity brought on by cold. Curiously, the birds that nest on the arctic tundras and wetlands of the Far North generally start south at the end of their nesting season; those from more temperate latitudes, especially if they depend upon insects or nectar as food, commonly fly to warm lands before autumn has reduced the availability of their usual nourishment. Within the tropics, the native birds cease breeding during northern winter because they cannot find food for extra mouths while so many competitors are present.

During the cold months in the North, both the brown (actually black) bears and the grizzlies retire to dens in which they will not be disturbed while sleeping and fasting until the plants resume growth in spring. Smaller mammals, such as marmots (groundhogs) and ground squirrels, go into true hibernation. They lose consciousness, their body temperature goes down and they live on their reserves of fat for as many as 8 to 10 months. Chipmunks and some rodents that store large quantities of seeds, dry fruits, and mushrooms in subterranean chambers alternate between sleeping and eating until a new growing season begins. Hibernating bats in caves, where the temperature never reaches the freezing point, awake at intervals; they fly back and forth, apparently testing the night air outside until they find edible insects.

Warm-blooded mammals and birds that remain exposed to low or high temperatures face the full challenge from this aspect of their physical environment. Their tolerance is broadened by the insulating effect of fur or feathers, which help in adjusting the rate at which heat leaves or reaches the body surface. Evaporative cooling, by expending water in sweat, is a surface phenomenon, just as is ordinary heat exchange with the environment. But heat production by muscles and glands occurs essentially in proportion to the bulk rather than to the surface area of the body.

Inherited patterns of growth show relationships to both temperature stress and the ratio between body weight and body surface area. As the German physiologist C. Bergmann pointed out in 1847, warm-blooded animals of the same body form tend to be larger in cold climates and smaller in warm ones (Bergmann's rule), undergoing an increase in the bulk of heat-producing tissues without a corresponding enlargment of heat-radiating surfaces. An example is the cougar (*Felis concolor*), which is the most wide-ranging species of mammal in the New World; it grows largest in British Columbia and Patagonia, and smallest in Central America. The emperor penguins, which breed on Antarctica in winter, are the largest of these flightless birds, and the Galápagos penguins—the most tropical—are among the smallest.

The British mammalogist J. A. Allen observed in 1877 that closely related species have proportionately less surface in cold climates than in warm through

Prior to regular seasons of drought or cold, deciduous trees and shrubs form winter buds that resist desiccation. Within the buds, which may be covered by waterproof bud scales, are very short stems that bear embryonic foliage or flower parts or both. When warm, wet weather arrives, the buds open and the stems lengthen, perhaps displaying reproductive organs before the foliage expands. Pendant catkins of hazelnut (top), like those of birch and alder, release pollen into the wind for dissemination; the pollen is captured by small, petal-lacking flowers upon the branches. Magnolia flower buds (bottom) open early in the season and attract insect pollinators before the leaves expand.

Ptarmigans (top) are found at high latitudes and high altitudes. Their feather-covered feet are well adapted to standing on snow and their plumage changes from brown in summer to white in winter. Cranes do not adapt to their winter environment but migrate to warmer climates. Japanese cranes may occasionally be forced to walk about on snow (bottom) after a spring storm.

reduction in the size of ears and tail, shorter legs and neck, and more compact body form (Allen's rule). Clear examples are found among rodents, rabbits and hares, foxes and wolves. But many exceptions have been discovered to both rules, such as the long neck of the polar bear and the giant size of the African elephant.

Temperature affects marine fishes and invertebrate animals directly, despite the mixing produced by currents and wave action and the thermal inertia of water. Among fish species that range from warm to cold waters, whether vertically in a temperature gradient or horizontally from equatorial to polar regions, individuals in cold waters show a greater average length of body and greater average number of vertebrae. Called Jordan's rule after its discoverer, the distinguished American ichthyologist David Starr Jordan, it reflects extra growth at a slower rate in low temperatures. Sexual maturity comes later too. Jordan further noticed that in cold waters, fishes of more species lay larger eggs with more yolk, or are ovoviviparous, or provide parental care for their eggs and young. A similar correlation between brood care and low temperatures has been discovered among the echinoderms of polar waters.

In fresh waters, the effect of temperature on life is mostly indirect, through altering the availability of oxygen. As water cools from 77°F (25°C) to the freezing point, it can dissolve about 40 per cent more of this important gas. At low temperatures, water offers more oxygen to gill-breathing animals when they are less active and offers less oxygen at higher temperatures when their demands increase. This difference in dissolved oxygen is critical for active predatory fishes in streams and rivers, such as trout, which survive so long as the water is shaded and kept cool by overhanging trees, but die and disappear if the trees are cut and the sun warms the water.

With diminishing temperature, the density of water increases. Anomalously it reaches a maximum at 39°F (4°C), rather than at the freezing point. It becomes abruptly about 9 per cent lighter as it freezes, due to the arrangement assumed by the ice crystals. These physical features of pure water are responsible for thermal stratification of deep ponds and most lakes, for periodic overturn of the whole lake when the water in it is all at approximately 39 degrees, and for the roof of ice that floats on fresh waters where the winter temperatures stay long below the freezing point. All of these contribute to natural selection, challenging the tolerances of living things, and determining ultimately which are well enough adapted to inhabit fresh water—flowing or still—at each level and every time of year.

Fishes sometimes are so sluggish in cold water that they let themselves be surrounded by the thickening roof of ice. Yet if their reserves of food maintain their metabolism and a slow release of heat, they ordinarily survive without freezing. Enclosed in a watery prison, their greatest dangers are chemical: scarcity of dissolved oxygen, and inadequacy of the reservoir for carbon dioxide and wastes from respiration and excretion.

Because of its dissolved solutes, sea water freezes in a different way. At about 29°F (−1.91°C), long slender crystals of pure ice separate and float. The sea water from which they formed may have been 3.5 per cent dissolved salts and 96.5 per cent water. But removal of the water to form the ice makes it more concentrated and denser. It sinks and is replaced by full-strength sea water next to the ice. As more

The behavior patterns by which animals avoid the winter are many. The chipmunk (below) collects a substantial store of dried fruits, seeds, mushrooms and other foods into underground chambers, then builds a soft bed of plant fibers atop one mass of food and curls up there to sleep. At intervals it awakens, relieves itself of wastes in another chamber, eats a meal, and goes back to sleep.

Ground hogs and other marmots store nothing but fat and hibernate while using up these reserves inside the skin. Monarch butterflies (right) migrate in swarms from Canada and northern states south to the Gulf States and Mexico. In spring, mated females fly north, laying eggs on young milkweed plants and recolonizing the distant range.

thickness of overlying earth and rock, show the most nearly constant temperatures on record. They vary in many instances less than a tenth of a degree year after year. Yet animals living so far from the sun and temperature changes respond to the seasons by timing their reproductive activities on an annual cycle according to variations in the oxygen and nutrient materials reaching them in their isolation.

The Limiting Effects of Pressure

The weight of overlying water increases the hydrostatic pressure in the sea by about half a pound per foot of depth. At the bottom of the Mariana Trench in the Pacific, the pressure reaches 9 tons per square inch. This pressure and the low temperature combine to make the water almost as viscous as molasses. To move through this medium, abyssal animals and those that scavenge over the sea floor at great depths need special adaptations for moving about and apparently do so slowly.

Pressure tends to stratify animal life, for each animal is immune to the compressing forces in its environment only so long as it stays where its internal pressures are approximately equal to those around it. But some kinds do make vertical migrations, either during their lifetimes or on a daily schedule, as a means of benefiting from the marine environment without succumbing to the hazards. These vertical migrants will be discussed in the chapter on the sea.

Variations in pressure in air, between sea level and the highest habitable regions on mountain slopes, influence both plants and animals in ways that are discussed in the chapter on high altitudes and cold regions.

ice forms, it tends to form a three-dimensional latticework in which pockets of sea water are captured. The concentration in these increases as further water turns to ice. At about 13°F (−8.2°C), the solubility of sodium sulfate may be exceeded, and crystals of it grow rapidly. At 10°F below zero (−23°C), sodium chloride crystals appear too. Depending on the rate of freezing, sea ice varies greatly in its content of captured salt, from about 1 per cent in new ice forming under an air temperature of 40 degrees below zero to almost none.

Cold, dense water from arctic and antarctic coasts often maintains its temperature between 35° and 30°F (+2° to −1°C) as it descends to the bottom and flows slowly into the great depths. Sites in the deep sea, like those in caves insulated by a great

The Rhythmic Tides

In tropical reefs, coastal shallows, and tidal gutters, the physical environment changes rhythmically according to the ebb and flow of tides. Burrowing bivalves, barnacles, and many other animals cease feeding when the water drains away. Fiddler crabs are equally limited to daylight hours when the ebbed tide allows them to scavenge over exposed sandbars and mudflats; at other hours and at night, these crustaceans hide in their burrows, usually under water.

Many annelid worms, crustaceans, mollusks, and fishes are similarly affected by tidal changes. Directly or indirectly their resources change according to the depth of the water, and hence to the pull of the moon on the earth, augmented or partly suppressed by the gravitational attraction of the sun for our planet. Their lives follow a lunar or semilunar periodicity, and often exhibit special features that correspond to the particularly great tidal changes (spring tides) at the full and the dark of the moon. Reproduction may be limited to times of spring tides.

Fiddler crabs show a regular interaction with both tides and daylight in the expansion and contraction of their pigment cells. The complex rhythm continues for almost a week under the constant conditions in a laboratory darkroom, combining lunar and solar timing. Thereafter irregularities develop, although the crab can quickly reset its clock from any variations in the pressure of sea water rising and falling and in the intensity of light.

The Limiting Effects of Light

Rhythmic cues reach most kinds of plants and animals from the daily rising and setting of the sun, and the changing length

of night throughout the year. Often their responses, called photoperiodism, can scarcely be attributed to light alone or directly, since the same pattern is followed by temperature, availability of water, and features of the chemical environment due to the activities of green plants and the effects of consumers and decomposers.

Except in the summer in high latitudes, each plant with chlorophyll goes through a daily diminishing of light intensity until, to stay alive, it must draw upon its reserves of energy and lose weight. Then the sky brightens until, at an intensity of illumination known as the compensation point, the production of organic material through photosynthesis exactly matches the plant's respiratory use of organic compounds in normal metabolism. The light brightens, and the plant can store a little—first toward the needs of the coming night and then toward growth and reproduction. Whenever the light intensity exceeds tenfold that at the compensation point, the efficiency of storage falls off because the extra energy destroys some products of photosynthesis.

Through a mechanism that has yet to be elucidated, many land plants respond to the length of night by flowering when nights are short ("long-day plants") near summer solstice or when nights are long ("short-day plants") near autumn equinox or even winter solstice. The flowering of dandelions is suppressed in mid-summer, whereas that of poinsettias comes normally in the tropics and the warm parts of the Northern Hemisphere as Christmas approaches. The flowering of a poinsettia plant can be prevented experimentally by shining a light on it for two minutes in the middle of the night. The action spectrum of this adaptive response to light matches fairly well the absorption of energy by chlorophylls. No known product of photosynthesis seems involved, and attempts to discover the chemical mediator (named florigen) have so far been fruitless. That the mechanism is important to plants seems clear, and that light (or its absence) is the limiting factor remains unquestioned.

The daily rhythmic change in light intensity is followed for a while even in total darkness when certain kinds of plants and animals are transferred out of their normal habitats. The leaflets of wood sorrel (*Oxalis*), like those of clover and many other members of the pea family, continue to droop before sunset and spread again before daybreak. In constant dim illumination, hawks and gray squirrels go to sleep each night and are active by day, whereas owls and flying squirrels follow the reverse schedule. Extra light at night may not affect the normal rhythm. Among the irregular projections of a tropical coral reef, the parrot fishes seem able to ignore a lamp shining on them as daylight fades. They seek out their favorite places for the night, and secrete about themselves a mucous envelope in which to spend the hours that normally are dark. Yet if transferred to an aquarium that is illuminated regularly at night and kept dark by day, they shift their schedule to match. In some zoos today, nocturnal animals can be seen by human visitors in special display buildings where the white lamps shine brightly all night, but only red lights (which the animals cannot see) glow all day.

The physical environment is affected by the presence and activities of living things. Each organism, no matter how small, releases heat, casts some shade, and exchanges wastes for raw materials. This feedback is evident in a forest with interlocking leafy branches, below which the temperature changes far less than above the trees. Wind may largely be excluded from the forest. Relative humidity is higher among the trees, and like the proportions of oxygen and carbon dioxide in the air under the forest canopy, shows measurably daily variations that match the physical changes in temperature and light.

In the surface waters of the seas and of rivers, the shading effect is due to myriad microscopic cells and small animals. The greater the number of individuals, the more turbid the liquid appears and the less the light penetrates to lower levels. Some springs and the Sargasso Sea, like small lakes high on mountain slopes, contain so little life that they are extraordinarily clear.

Sometimes we become aware of the heat that living things produce. A gaggle of geese swimming all night in one small part of a pond or estuary can keep the water there too warm to freeze and trap them. In a hollow tree, with only a thickness of bark and wood as insulation keeping out the cold wind, a colony of honeybees can maintain an average temperature in the 50°'s Fahrenheit so long as their reserve of honey furnishes the metabolic fuel.

Heat, light, and the currents of air and water powered by solar energy in great convective eddies, all combine in our concept of climate. The currents redistribute heat and moisture, and disperse hordes of living things, particularly those of small size, which are the most abundant of all. Yet in a synthetic mode of thought we must fit the physical and the chemical aspects of the environment into a matrix like a net. It restrains living things, while yielding to their presence. With the facts learned through analysis fitted together properly, we can visualize the matrix as a whole. We call it the environment of life.

6 Avoiding conflict within the species

Charles Darwin recognized that "competition should be most severe between allied forms, which fill nearly the same place in the economy of nature." In the third chapter of *The Origin of Species*, he gave his opinion that "the struggle will generally be more severe between species of the same genus, when they come into competition with each other, than between species of distinct genera." Members of the same species have the most conflicting interests of all, since they compete for the resources in a single ecological niche. As the population grows, the resources available to each individual diminish. Competition becomes severe, even if aggressive action can be avoided. The rarity of overt conflict between individuals of the same age and species points to the evolution of a multitude of different mechanisms whereby the struggle for survival is quieted. Avoiding conflict has become an art.

In many instances among both plants and animals, the resources in the immediate vicinity of the reproducing individuals are kept for their use by having their offspring disperse quickly. An additional advantage can be seen if the young occupy a different habitat, one for which they are specially adapted and where they will be least likely to compete with mature individuals of their kind. The clearest examples live in the marine environment at moderate depths, where the parents may inhabit the bottom or the shores while the larval forms move among the drifting life in surface waters.

Some of the stationary plants, whose disseminules—new individuals being dispersed—travel no great distance, show adaptive features that relate to the different microhabitat and microclimate shared by young individuals. The sprouts that rise from the roots of an oak tree and many seedlings two to six feet tall bear enormous leaves in which their chlorophyll is spread to the dim light under the forest canopy. On tall trees of the same kinds, the high foliage receives more sun and each leaf is smaller and lets light reach those on the lower limbs. In the almost shadeless forests of gum (*Eucalyptus*) trees in Australia, the sunlight that streams down between the narrow pendant leaves on high branches often strikes bluish gray juvenile foliage on young shoots near the ground. The young leaves commonly are as broad as long and arise in successive pairs along the stem. These differences affect also the animals that are associated with the leaves at various levels in the forest.

The time of life is sometimes more important than the time of year in determining the relationships between a plant or animal and its environment. Both change meaningfully from week to week, the one in a lifetime and the other cycling through the seasons. These changes are often most significant in minimizing competition for resources between immature and adult members of the same species.

Generally a correlation can be found between the number and size of the disseminules being dispersed in a generation and their freedom to be carried passively by currents of air or water. Yet even the fewer, larger young of shorebirds and waterfowl commonly fly from the nest area as soon as they are independent. Their directions seem to be at random, although the survivors among these birds will later travel with others or navigate on their own along the migratory routes of their species.

Among the disseminules, mortality is usually high. The sporelings and seedlings of plants, like the larvae and juveniles of animals, have a meager store of energy on which to survive any inclement weather. At this age they are least tolerant of a deficiency in any essential nutrient and most vulnerable to diseases and predators. Yet even when many small young individuals are present in a limited space, their collective impact upon the environment may be too slight to recognize as competition.

Each species is particularly well served if the dispersal that is normal in reducing competition in the home area gets a few disseminules to a distant region that proves hospitable. Slow changes in geological features do provide opportunities for pioneers that can cross customary barriers, no matter how infrequently. Often in the new situation the pioneers find a tolerable, although unfamiliar combination of resources. If the fresh arrivals succeed in colonizing the area, they may quickly occupy a number of unlike niches. If isolated to this degree, the colonists can become the ancestors of separate species, each defined and molded by its niche. Speciation—the formation of new species— seems to occur most commonly around the fringes of the geographic range of the parent type.

Beneficial Neighbors

Nearness to other individuals of the same species does not always imply intense competition. Sometimes it shows merely that the habitat or the way of life meets the needs of very few species and these tend to grow side by side. In other situations, appropriate adaptations do help organisms of a single kind gain by togetherness.

When plants of the same kind are close together, wind can easily cross-pollinate them. This is noticeable among grasses on

grasslands, and at high latitudes and high altitudes where the flora is simple. Dense clumps of grass, however, arise by vegetative reproduction rather than by seeds. Solid stands of fir and spruce, which are usual in northern coniferous forests, benefit by nearness in wind-pollination and in having the shallow roots of one tree intertwined with those of the next, preventing gales from toppling them. Toward the Equator, by contrast, solid stands of trees and wind-pollination are rare, and so is competition among plants of a single species. Generally tropical trees of each kind are so distant from one another that cross-pollination is possible only through the unwitting assistance of winged animals.

Among animals, aggregations that might be compared to solid stands of plants

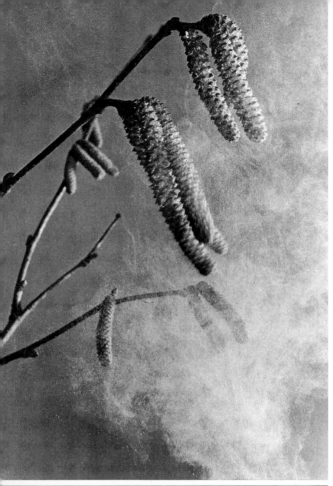

are less frequent. If individuals of a species come together with a minimum of conflict to feed side by side or to remain dormant, the gain seems to be through greater safety from predators (hence other species) rather than through any lessening of competition for limited resources. Yet instances do stand out in which animals benefit from company while exploiting their environment. More often the individuals remain well dispersed during most of each life cycle, reducing competition, but assemble in numbers for reproduction. Or adaptive social behavior may regulate the size of the population according to the amount of suitable habitat available. Each of these ways of life can be found well developed among familiar species. Generally a study of the aggregations reveals the gain.

Occasionally the benefit from companions is easy to assess. A tangle of serpent stars (ophiuroid echinoderms) on the sea floor seems far more successful in capturing food particles from the current that flows among their arms than any solitary individual of this kind. A cluster of the common marine flatworms (*Procerodes ulvae*) can survive in a tide pool whose salinity has been reduced by a rain shower, where single worms of this kind perish. The clustered worms, each about 1/4 inch long, release calcium-ion in their immediate vicinity as an emergency action that reduces the osmotic stress until the tide returns and ends it. The isolated worm cannot sufficiently influence its chemical environment.

Scientists who use fishes as test animals in laboratory tanks to measure the effects of pollution have learned the importance of population size. Fishes in a group at modest density may survive the addition of colloidal metals because the mucus over their scales precipitates the poison. In water polluted to the same concentration, one fish or a few cannot reduce the concentration of the toxic material enough to make it tolerable, and they succumb quickly.

Geneticists know that fruitflies (*Drosophila*) in the isolation of a culture jar show a longer survival time for individuals if there are from 35 to 55 flies per one-ounce container than if there are either fewer or more. These insects eat the yeast fungi with which the nutrient medium in the jars is inoculated. With fewer flies, the yeasts proliferate and use up the nutrients; they die, and let the flies starve. With more flies the jar becomes contaminated with feces and the yeast cells are cropped so severely that their reproduction cannot keep up; soon the flies are starving. The right number of flies devour the cells at a rate that promotes continuous growth.

Breeding Aggregations

Temporary aggregations that match the reproductive season are well known in many animals. Their regularity has occasioned much comment, and stimulated many biologists to search for periodic cues from the physical environment that can trigger the biological phenomena that recur on schedule. The study is called phenology, and reveals that the timing by which each species lives leaves little leeway. Preparations for mating and egg-laying that come several weeks early or late are often futile. Natural selection normally marks these individuals for oblivion. The future belongs to the conformists, which follow closely the pattern that has been incorporated into the genetic heritage. Readiness on schedule is no coincidence. It has brought success to previous generations. Generally readiness and nearness to others of the same species are both essential.

Behavioral ecologists are still trying to identify the combination of cues from the environment that induce certain kinds of marine fishes to approach the coast simultaneously in such astounding numbers. The grunions along the California shoreline, the capelins near Greenland, and the alewives in New England estuaries gather within sight of man and many predators. Maneuvering, meeting, and mating, these fishes multiply their kind, then return quickly to the offshore waters in which they can disperse while finding food and shelter.

Similar behavior is shown by horseshoe crabs (*Limulus*) and by many of the marine worms. The segmented nereids especially, propelled by their rows of paired paddles, swim to the sea surface and engage in mating ballets, one species at a time on an inherited program of immense antiquity. Each species shows its preference for certain months and times of day or night. But usually these activities that produce aggregations at mating season match the tidal rhythm and in it the full or the dark of the moon.

With equal regularity, amphibians and water birds come to the edges of fresh water wetlands to pair and procreate. Some of the individuals that gather have just reached their sexual maturity. Along with older and presumably experienced individuals, they assemble into impressive aggregations, only to disperse as soon as the reproductive activities have been completed.

Sexual activity and care of the young both rank high in the spectrum of features that reward animals for relinquishing their dispersed condition and forming periodic aggregations. Yet much of the travel undertaken by each individual precedes any communication by sight or sound or scent

69

Survival in Groups

Where a natural resource is locally abundant, animals often congregate in large numbers. Rabbits in Australia (top, right) come by the dozens to drink from ponds in semiarid land, just as fishes swim in large schools (bottom, right) if nourishing plankton is abundant in the aquatic environment. Flying foxes, which are fruit bats with a wingspan of 3 feet or more, cluster by day in Burmese swamps (top, far right) and then disperse at night in search of food. The marsupial koalas of Australia, by contrast, form small family aggregations, mothers carrying their young on their backs (bottom, far right), until each is weaned and able to choose edible foliage from the few suitable kinds of eucalyptus ("gum") trees.

from members of the same species that have already reached the common destination. Salamanders creep silently across the countryside, sometimes from more than a mile away, to their home streams and ponds. Frogs and toads that have arrived are noisier, calling individually or in chorus, beckoning others of their kind to join the throng engaged in promiscuous copulation (called amplexus) and to place eggs where the hatching tadpoles will have a suitable habitat.

The distance traveled to reach a breeding aggregation after wide dispersal proves the advantages of minimizing competition for resources during most of the year, and of regular return to a part of the world where the young will have food and suitable shelter. In making these journeys, which are often fantastic, each individual relies upon its own homing ability. This is a combination of inherited guidance toward a general geographical destination, and of experience with physiographic features in the individual's place of origin.

The extreme travelers are the arctic terns that flit over the open oceans, plunging shallowly for small fishes near the surface whenever they are hungry, from near the North Pole in northern summer to the waters around Antarctica six months later. Almost as spectacular are some of the shearwaters that nest on islands near the southeast corner of Australia. Each fledgling to become airborne begins a clockwise circuit around the whole of the Pacific world, flying and feeding on a course that brings it back to its hatching grounds. Despite their dispersal during most of the year, 95 per cent of these nesting birds arrive within a three-day period, and the rest straggle in before the week ends.

Often a large number of migrants converge on a small area, and nest side by side. Yet each cormorant, gull, gannet or penguin actually defends a small territory around its eggs or young—generally as far as the parent's beak will reach while incubating or brooding. Each parent feeds only its own chicks. Similarly in a rookery of seals, the big male defends his harem from other males, but each female tends her own pup without help. Every cow elk in the harem guarded by a bull protects and feeds her own calf and no other. This is tolerance of near neighbors rather than social behavior, since no parent helps except its mate.

The parent waterfowl, the mother seal, and the cow elk maintain themselves by frequent feeding, and share the nourishment they get with their young. But the bull seal fasts throughout the many weeks during which he is protecting his harem and the females their nursing young. Most of the amphibians and fishes eat nothing while on their reproductive territory. The only competition is for mates.

Social Aggregations

True social cooperation is most obvious among certain insects, such as some wasps and bees, some ants and termites, and the various primates. Analogous patterns of behavior can be recognized in those colonial coelenterates in which each feeding or reproducing branch appears to be an individual, although all may share the same digestive cavity. The coelenterates seem much less complex because their nervous coordination is so elemental. Yet their parts are in continuous contact and attend to the operations of the whole through a remarkable division of labor and responsibility. Social aggregations of more complex animals may include division of labor too.

But the social grouping differs from a mere breeding aggregation in that care of the young is shared by adult females other than the actual mothers, and sometimes by adult males and juveniles.

The nests of social insects are generally dark inside, which restricts the cues between the separate cooperating individuals to chemical, tactile, and audible signals. The white-faced hornets (*Vespa maculata*) build their carton nest of chewed wood pulp in early spring, suspending it below a tree branch as the brood shelter for eggs and young. These develop through the grub and pupal stages to become fertile adults in about a month. Progressively the mature females enlarge the nest until, in autumn, it may be the size of a bushel basket, housing 15,000 individuals as well as the immature progeny they guard and serve. By contrast, female bumblebees build their honeypots and brood chambers in groups openly on the ground, where a few adults remain in attendance to drive off invaders of other species.

Honeybees, which differ in having a worker caste of sterile females in addition to the fertile queen, readily revert to their ancestral habit of building their honey and brood comb inside hollow trees. Eventually any prosperous colony of these insects becomes overcrowded, and a swarm of as many as 60,000 workers leave with the old queen and a few drones (males) as hangers-on. A swarm is an impressive aggregation of individuals, extraordinarily specialized in that almost all are matured nonreproductive offspring that remain in attendance upon their mother and assist her in launching new generations. The queen lives for several years, laying eggs daily all through each summer, relying upon workers to build the comb, care for

the eggs and larvae, gather and store the food, and defend the colony against disturbance by animals as diverse as other bees and hungry bears. Each honeybee takes about 16 days to reach maturity, and the turnover in the colony is continuous because, unlike the long-lived queen mother, the workers and drones (males) survive less than four weeks of activity.

The only conflict within the honeybee colony is one-sided and clearly adaptive. After the population and its store of food and developing young reaches a critical point, the old queen and a swarm of workers move out to found a new colony. In a day or so, one of the new queens emerges and promptly explores the hive, stinging to death every other developing queen. Only then can she go off with a following of eager drones on her mating flight, secure in having a well-stocked home to return to after she has accumulated her lifetime supply of sperm cells.

Colonies of ants and of termites ("white ants") seem special because the caste system has been extended to include big-jawed soldiers as well as workers, and sometimes others, such as the food-storing repletes and the acid-spraying nasutes. Termites have direct development, hence no pupal stage, and it is the immature individuals that do most of the work in all except the most advanced families. (By contrast, among ants, only the adults perform the duties of the colony because the young are maggot-like and the pupae are enclosed in silken cocoons resembling puffed rice, mistakenly called "ant eggs.") Termite workers eat the food, share the products of digestion, clean the galleries and extend them as dark, humid runways to supplies of nourishment. Among the social hormones shared from termite to termite

Social Organization

Social insects exhibit a range of adaptive relationships that promote the safety and rapid development of young individuals in each colony. The long-lived queen of the bald-faced hornet (Vespa maculata) *survives the winter and, in the early spring, starts a nest, in which all the first brood of offspring will mature as worker females. They enlarge the nest with chewed wood fibers, creating a paper nursery (top, right) in which, by autumn, the number of cells may exceed a thousand. During the hornet's metamorphosis many of these cells are used several times in sequence as repositories for an egg, for the maggot that hatches out, for the pupa into which the full-grown maggot transforms, and, finally, for another adult. Workers store no honey or pollen but feed the queen and the maggots daily on regurgitated animal matter such as caterpillars. The workers can lay eggs which develop into males called drones. In the autumn, the drones mate with the young fertile females and it is these that hide as best they can from winter.*

Termites, often called white ants, live where winter is less of a challenge and in the tropics often build nursery nests well above the ground, such as among the spreading roots of a screw pine (bottom, right). The young that hatch from eggs laid by the queen or queens are active and serve as workers regardless of whether they mature as sterile soldiers, some other caste, or as fertile males or females. All active individuals in the colony share in the regurgitated food offered by workers that gather it. All workers cooperate in keeping the walls and covered runways of the nest closed off from outside air, which may be intolerably dry at times, and from ants.

The comb produced by honeybees (far right), by contrast, serves both as a nursery for the helpless developing young in individual cells and also for storage of honey, pollen (in the form of "bee bread"), and sometimes water, for use of the offspring of the single queen. With a modest amount of shelter as insulation, the hive can be kept warm in winter by the activities of the clustered workers who protect the queen at the center of the group and draw upon the honey stores for energy. Worker honeybees, who comprise virtually all of the adults in a colony, follow a program of changing duties within the hive from the time of their emergence from the pupal chamber until they go outdoors—first as door guards and fanners to ventilate the hive, then as field bees to collect food until their wings wear off from use.

When threatening gestures do not establish dominance between two males of comparable size, combat may begin. Young males spar in practice, but old ones battle to retain their right to a harem. Bloody contests occasionally develop among full-grown American bison on the western plains (left). When two sable antelope bulls go down on their knees, facing one another on an African savanna (lower left), each can estimate the opponent's size; a few inches difference in the length of the curving horns may settle the dispute. In alpine meadows of Europe, ibex goats (bottom right) often jab at one another with apparent fury until one accepts an inferior position in the hierarchy and runs off. On the western grasslands of America, male sage grouse (below) confront each other, inflating enormous neck pouches and spreading spiky tail feathers as the first, and often the only, act needed to establish one as the dominant bird in the area.

must be some from the sexually active "king" and "queen" that inhibit the maturation of other termites that could be fertile. When accident or old age removes the current egg-layer or her constant consort (who is no drone!), new reproductive individuals mature to take over the procreative role in the termite colony.

The cooperative behavior of individual social insects precludes competition among the members of each colony by regulating its numbers according to the availability of resources in the environment. Among social primates, the size of the population is equally regulated in relation to the amenities of the habitat by a flexible hierarchy of relative dominance. Both males and females separately engage in frequent confrontations, but settle—more often by bluff than by battle—which of any two individuals has the right to first choice. The dominant male has first access to mature, ready females. If he shows a preference for one female, as is often the case, she gains corresponding status—if she did not have it already. When the dominant male tires, as sometimes happens, a number two male is ready to take over and is allowed to do so. These relationships are tested at frequent intervals by grimaces and gestures, and emphasized by calls of many kinds. Communication is effective both in excluding individuals of low status from many of the available resources, and also in alerting the troupe to any youngster that gets separated. The young individual can summon help, which often includes a high-ranking male or two, the mother and a few "aunts."

Competition for Territory

The British ornithologist H. Eliot Howard in 1920 pointed out the significance of territory in the lives of many animals. He observed that the males of many different species of song birds space themselves in suitable nesting habitat, sing their distinctive calls at intervals as an audible claim, and seemingly accept as a mate whatever female of their kind chooses to stay and build a nest. Until their young are fledged and dispersed, the pair drive interlopers of the same species from their territory.

Actually almost any small bird or medium-sized one is treated as a trespasser close to the nest. But the vigor of pursuit diminishes in proportion to the distance from the nest itself. At the boundary of the territory, the claimants on both sides make equal protest; this equality is the essence of the dividing line.

The pre-dawn chorus of song birds in nesting season serves each species well even before the sky brightens enough to make food visible. Each resident male moves to his singing perch, and begins to warn off intruders of his kind with almost the regularity of a foghorn. Any newcomer of the same species entering this subdivided real estate is ejected promptly, by the resident male if a male, and by his mate if a female. Rarely does a late arriver have sufficient vigor to oust the bird already in possession.

Now that biologists have found ways to mark animals distinctively and to recognize the same individual wherever it is seen at close range or recaptured, the extent of territoriality in the world is becoming better appreciated. Previously it was hunter's lore alone that gave a cougar the reputation of defending the area around the hill where it made its lair, and of indicating the boundaries with little piles of fallen leaves wetted with an odorous urine. The Asiatic tiger is known to patrol a territory in the same way.

On West Indian islands, the small lizards (*Anolis*) every day take up positions from which they can dash after and frighten away any member of their own species and sex that comes into view. Rarely does the intruder have a chance to eat any of the insects and worms that are the chief amenities in the area. Among the irregularities of a coral reef, individual crabs, octopuses, and fishes recognize boundaries and stay in a home range of surprisingly restricted dimensions.

Territoriality tends to isolate local groups and to favor inbreeding. Each group can become an unrepresentative sample of the full genetic heritage of the species, showing a provincialism in its inheritance to which the name "genetic drift" is given. Often this is the first step toward subdivision of a single species into several. If changes in the physical environment confer survival value on further steps, the evolutionary process may continue and make the new species distinct.

Speciation

The natural processes that lead to separation of new species are known collectively as speciation. In the decades since Charles Darwin and others drew attention to evidence indicating that modern species have evolved from fewer in the past, the biologist has discovered far more criteria than formerly upon which to decide whether two individuals are of the same or separate species. Members of a single species are expected to show uniformity in fine chemical details, in the sequence of genes in their chromosomes as well as in the shape and number per cell of these carriers of inheritance. Genetic isolation through geographic or behavioral separation, which reflects some uniformity in ecological niche, is anticipated. Different species occupy unlike niches, and may be sufficiently distinct to show impaired fertility in test crosses, such as that between a horse (*Equus caballus*) and a donkey (*E. asinus*).

The Asiatic tiger (*Panthera tigris*) has long been isolated geographically and behaviorally from the lion (*P. leo*). Even when the range of the lion extended beyond Africa south of the great deserts into North Africa, southern Europe, Asia Minor and across to India, this species with its maned males kept to savannas and open country where it had little chance of contact with the tigers of swamps and wetlands. Yet, in a zoo, a lion will mate with a tigress or a tiger with a lioness and produce a hybrid, called a liger. Neither lion nor tiger will respond to a leopard (*P. pardus*), a forest animal of Eurasia and Africa whose geographic range is close enough to wild lions and wild tigers for hybrids to arise if the behavioral differences did not serve to isolate the species.

Ecological research reveals the great diversity of the habitat occupied by the members of a single species population over its total geographical range. Some behavioral differences are necessary to cope with differences in climate, in soil, in the abundance of particular parasites and predators, in the supply of food; hence the density of the species population and competition within the species. All of these are factors in the resistance of the environment toward growth of the population. They provide the forces that mold the evolution of each local subpopulation. And they change measurably throughout each season, from year to year, and from decade to decade.

Territorial Imperative

Possession of a territory is a prior requirement for successful reproduction among many kinds of animals. The individual who can defend the boundaries is commonly the most vigorous male; his offspring may have the best opportunity to develop because of an abundance of food and a minimum of molestation. Even the Dotilla crabs whose young spend their early weeks of growth among the plankton will defend the igloo-like mounds the males build upon the shore (far left, top). Contests between male Canada geese (far left, bottom) tend to be more ritualized, while potential mates stand by. The timber wolf (left) uses his loud voice to claim for his family all territory within earshot, while the red fox silently leaves scent posts (left, below) marked with odorous urine in which crystals often glisten.

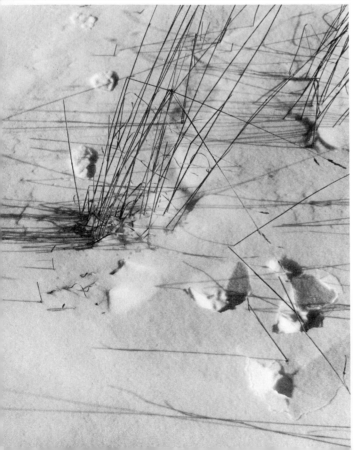

The chart, below, shows the ecological niches of ten of Darwin's finches on one of the Galápagos Islands. The woodpecker finch (right) is unique among these finches in being able to climb vertically on trees and is unusual among animals in being able to use a tool. After pecking a cavity in the bark and exposing an insect larva, the bird plucks either a stiff spine from an Opuntia cactus, or a twig of similar size, and holds it lengthwise in its beak. Probing with this tool, the bird extracts the insect larva from the cavity. The same behavior is shown on all seven of the islands inhabited by this species.

Geospiza magniros	*Geospiza fortis*	*Geospiza fuliginosa*	*Geospiza difficilis*	*Geospiza scandens*	*Platyspiza carssirosus*	*Camarhyn. psittacula*	*Camarhyn. parvulus*	*Cactospiza pallida*	*Certhidia olivacea*
big seed ground finch	medium seed ground finch	small seed ground finch	scratching ground finch	cactus finch	vegetarian tree finch	insectivorous tree finch		woodpecker finch	warbler finch
in Arid Coastal Areas			in Humid Forests	in Cactus Trees	on twigs, foliage, buds, bark			on branches, trunks	in mid-air
crushes food with edges of the beak					siezes food with tip of beak				snatches food in beak
eats mainly plant foods						eats mainly animal foods			eats entirely animal food

At the center of its range, a species tends to degenerate instead of to evolve. The most vigorous individuals that find no place to settle travel toward the boundaries as emigrants, leaving the successful conservatives to perpetuate the old ways until some invader—usually an unrelated species—supplants them. Often at the center of its range, the species may not be particularly numerous. Mature individuals may be uncommon. Yet at some moment in the life of each one is a critical stage when the natural resources will not support a larger number. The ecologist hunts for these critical points, for it is there that the environmental resistance suppresses the natural biotic potential for increase in each species. The extent of this regulation can be seen in the fact that, in any area, half of the plants and of the animals present belong to less than a tenth of the species found in the region.

In searching for the critical factors, the ecologist keeps in mind the heterogeneity of the genetic heritage within the breeding population. It is a normal diversity that arises because no individual has *all* of the beneficial genes. Nor does each beneficial gene provide its help at the same stage in the life cycle. Some individuals will be best adapted to a common situation met in the month of May, but find stiff competition in August from other individuals of the same age that barely survived the May conditions; now these survivors may be the ones most handsomely adapted.

A beneficial gene may relate to environmental conditions that are improbable. Inheritance can handicap some individuals for life at the northern end of the species range, but not at the southern—where they chance not to be. A progressive change in the weather will give new significance to this heritage. The possessors of the gene are often described as being "preadapted" to challenges that never arise in their lifetimes. Yet their descendants with this gene stand to benefit, perhaps in a decade, a century, or a millennium.

Around the edges of its range, each species tests its genetic variability against the tolerances needed to remain free of stress despite critical resistance from the environment. A new mutation or some beneficial combination of genes already present may let a few individuals and their offspring tolerate a feature of the habitat

that remains limiting, stress-inducing for the rest of the species. The favored ones multiply much faster and may replace the ancestral type. Evolution is in progress.

Genetic changes that temporarily reduce competition within a species by allowing some members of the population to occupy a slightly different ecological niche become most spectacular among the descendants of a colonist that reaches a remote new island. Because the variety of other species is small and so are the combinations of resources that prove acceptable, the survival of the colonists depends upon adopting new ways of life. In New Zealand, for example, the introduced house mouse has occupied seven different niches, only one of which is indoors. In Britain, by contrast, it is strictly a household animal and the six other niches it might have used are fully filled by native mammals.

The finches that Charles Darwin discovered on the Galápagos Islands could presumably have remained unchanged, like the original ancestor that arrived across the 600 miles of open water from the Ecuadorian coast. The population of descendants would then have been limited by the small assortment of seeds in acceptable types and sizes. Instead, natural selection favored an adaptive radiation of beak form and dietary habits to make use of these foods and others. The finches diversified into many niches that scarcely overlapped. Today, the four ground finches include one that scratches for seeds among the fallen leaves in humid forest, as a thrush might do. Three others in coastal arid areas feed on big seeds, middle-sized seeds, and small seeds when not hunting insects to feed their young. Cactus finches (*Geospiza scandens* on nine islands and *G. conirostris* on three others) eat the soft pulp of the prickly pear cactus and probe its flowers for nectar. The vegetarian tree-finch feeds primarily on buds, leaves, blossoms, fruits, and some kinds of seeds, generally in trees. Insectivorous tree-finches (*Camarhynchus psittacula*, *C. pauper*, and *C. parvulus*) eat medium-sized beetles and other insects found on twigs, bark and foliage; the ranges of these birds are centered on different islands, and overlap on only one. The woodpecker finch digs holes in tree cactus to reach boring insects, then uses a cactus spine to get out the insect. The nearest relative, the mangrove-finch (*Cactospiza heliobates*), feeds on insects but stays near the coast in the mangrove thickets remote from the other species. A warbler-finch darts after flying insects as though it were a flycatcher, and also hunts on trees for small insects, generally eating those with bodies softer than a beetle. The ecologist sees many niches instead of one, and regards niche separation due to change in habit and habitat as the first evolutionary step toward lessening confrontations and conflicts, and then toward speciation.

A cluster of islands, such as the Galápagos and the Hawaiian chains, favors the temporary lessening of competition through speciation because each new adaptive pattern can evolve and get established on a separate island. Later, when the population is big enough to foster emigration, the species can spread to adjacent areas. If it has achieved genetic isolation by then, it is unlikely to melt into the ancestral type by cross-breeding.

Probably most species have arisen in this way, with some measure of geographic isolation. Ecologists and evolutionists refer to the isolated population as being allopatric ("another country"). Only a few exceptions are known in which populations living in the same area (hence "sympatric") achieved the necessary separation of breeding stock to concentrate the new genetic combinations. One group may have become nocturnal while the parent stock remained active by day. Or a group began breeding in the spring, while the pattern of the ancestral type continues to require mating in the autumn. Matching the change is a separation of an ecological niche, with unlike requirements in space and food, offering a respite from conflict while serving as the prelude to speciation.

The ecological opportunities that allow living things to diversify have arisen in the past both through geological changes and also adaptive modifications in the structure and function of the plants and animals themselves. The fossil record proves that modern life occupies far more of the planet than it did 600 million years ago. The ecosphere has spread. The first big move seems to have come about 500 million years into the past, when living things spread from the seas into fresh water during Ordovician times. The colonization of water-soaked earth in the late Silurian or early Devonian (about 425 million years

ago), the evolution of somewhat drier forests during the Age of Reptiles (which began about 230 million years ago), and of the first grasslands in the Eocene (about 60 million years before the present) were pioneering changes with a great future. They permitted the total number of species to rise with the increased opportunities.

If the only major basis for diversification of ecological niches were the colonization of previously uninhabited regions of the world, then the number of kinds of life on earth might have reached its peak during the Pliocene when the deserts became populated. Some paleoecologists do credit a progressive decrease in variety during the subsequent two million years of the Ice Age, continuing to the present day. Others blame human exploitation of the natural world for the extinctions that have occurred during this period.

Extinctions that took place prior to the arrival of mankind on the scene must have made niches available to species that could occupy them. Whether the new occupant displaced the old or merely moved in where competition was lessened, the species in the niche changed without necessarily altering the number of niches. Presumably the birds and mammals inherited the niches of Mesozoic reptiles. Paleontologists have tried to estimate the extent of this turnover since the beginning of life on earth. They suggest that the fossils identified so far represent about one per cent of the living things that had fossilizable hard parts, and hence that about 500 million species have existed altogether in the past. If, as has been estimated, the total number of living species is close to 5 million (not just the 1.5 million that have already been given names), then about one per cent of all the species there have ever been are now alive.

7 How different species live together

Each person, like every animal or plant, lives in a community composed of many species. Whether these associates are harmful, harmless, or genuinely beneficial cannot be determined without careful examination. Sometimes the social relationship is evident from knowing what organic and inorganic materials the organisms produce or need for life. Ecologists examine carefully each environmental effect produced by one species that may facilitate or obstruct the constructive transfer of energy in each neighboring species.

Since each of any two species that interact may be harmed, or unaffected, or benefitted, there are six possible combinations in any relationship. We are most familiar with the situation in which both species suffer when together because both use the same resource. This is simple competition between species. We can easily recognize that one species benefits and the other suffers when a predator kills and eats its prey (predation) or a parasite takes nourishment from its host (parasit-

ism). We are less aware of the species that fend off another from sharing the same resources (amensalism) or that share resources more or less voluntarily (commensalism). Casual and nonessential assistance, with benefits, of one species by another, is protocooperation; whereas, an association that benefits both and is essential for the survival of at least one is called mutualism. In the final association, neither species is affected by the presence of the other; ecologists call the relationship neutralism.

Originally, all of these ways in which unlike neighbors shared the world were included in the concept of symbiosis—living together. The word was coined in 1879 by the distinguished German botanist Heinrich A. De Bary, who referred to it as "the appearance of cohabitation of unlike organisms." Today, in distinguishing the various forms of symbiotic relationship, ecologists rely increasingly on simple symbols, shown in the table below. An (0) can indicate a state of complete indifference,

The possible interactions between two different species

Relationship	Examples	Together	Apart
Competition	hole-nesting tree swallow	−	0
	hole-nesting bluebird	−	0
Neutralism	robin	0	0
	woodpecker	0	0
Amensalism	poisonous plant	0	0
	herbivorous animal	−	0
Predation	cat	+	−
	mouse	−	0
Parasitism	tapeworm	+	−
	cat	−	0
Commensalism	burrowing owl	+	−
	prairie dog	0	0
Protocooperation	oxpecker	+	0
	elephant	+	0
Mutualism	fungus of lichen	+	−
	alga of lichen	+	−

hence neither gain nor loss. A (+) shows a gain from the association. A (−) tells that the species loses, either from being close to the other species or from being forced away from its associate. To the ecologist, herbivorous and carnivorous habits are essentially alike in that the consumer eats part or all of a member of another species.

Competition

According to the usual explanation of evolution through natural selection, a slight advantage possessed by members of one species that compete for the same resources with members of a second will progressively allow one population to increase its consumption while the second loses out. For one species to replace another, the nature of the advantage is of little con-sequence. It may be a higher biotic potential, or a greater tolerance for some critical factor such as a periodic shortage of food, water, or weather suitable for activity. All of these factors can be included in estimating the fitness of a species for its habitat.

Ecologists have sought to measure competition. An indication of what they might expect to find could be gained from a series of mathematical formulations proposed independently by A. J. Lotka in the United States (1925) and V. Volterra in Italy (1926). A Russian-American biologist G. F. Gause decided to test these models experimentally, and thereby earned a position of special respect among ecologists. To simplify his analysis, he chose two species of slipper animalcules (*Paramecium*) whose food consists almost exclusively of hay bacteria (*Bacillus subtilis*). Gause recorded the changes in population size of each species as the animalcules reproduced in separate culture tubes and also as they competed in a single tube. Separately each showed the ordinary type of population curve, levelling off at the carrying capacity of the deliberately restricted environment.

When *Paramecium aurelia* and *P. caudatum* were started off in equal numbers in the same tube, *P. aurelia* always replaced its competitor, seemingly because of its slightly higher rate of reproduction. When *P. caudatum* was cultured with a third species (*P. bursaria*), however, both survived and reached stability. This proved to be due to segregation of niches, with *P. bursaria* feeding along the bottom of the container and *P. caudatum* eating bacteria in suspension. Extinction of one seemed inevitable if both food and space were realms of competition. Coexistence was possible if the niches differed in any way.

Interactions between members of different species follow many distinctive patterns. In salt marshes along the coast from Maryland to Texas, the saltwater cordgrass (Spartina alterniflora) *competes with dark clumps of needle rush* (Juncus roemerianus), *displacing its competitor where the salinity of the ground water is too high for the rush (far left). The great horned owl (left) hunts chiefly at night, preying* mostly on rabbits and rodents but taking an occasional skunk as well; this owl rarely attacks birds, but many birds, like this mockingbird, will harass the big owl if they find it by day. In Africa, mixed herds of zebras and brindled gnus (below) gain by associating together, the gnus having a better sense of smell but only fair vision, and the zebras having excellent sight but a poor nose for detecting predators near by.

To avoid having this discovery formulated into a scientific law or attributed to Gause or to any other single biologist, since so many contributed to its recognition, Garrett Hardin proposed to call it the "competitive exclusion principle." Briefly, it states that "complete competitors cannot coexist." Under natural conditions, complete competitors rarely exist at all. One species has a slight advantage over another in one area or in respect to one critical factor, and the other over the one in some different area or respect. On the same rolling American prairie, blue grama grass (*Bouteloua oligostachys*) carpets the crests, and little bluestem (*Andropogon scoparius*) the bottoms. Halfway between, more or less, they compete and overlap. In a series of dry years, blue grama replaces little bluestem on the slopes, whereas in a series of wet years, little bluestem grows almost to the crests. Each reproduces most effectively where its tolerances match the environmental conditions, particularly moisture. Each competes in a zone of overlap, expanding or contracting its total range according to climatic variations.

The principle of competitive exclusion is based on the assumption that the genetic constitution of the competing species will remain fixed. This too is rare. As soon as one species becomes dominant, competition between the species diminishes and natural selection operates in relation to contests within the dominant species for energy and space. By contrast, the dominated species is now little affected by competition within its population; for it, natural selection is mostly on differences between species. As mutations and recombinations continue in the genetic stock of the competitors, adaptive changes accumulate in the dominated species, helping it improve its competitive relationship with the other. Its population grows, even to the status of dominance. The likelihood of alternating periods of dominance is increased because the environmental conditions, such as weather, are never uniform from month to month or year to year.

Neutralism

The ecologist recognizes as neutralism the relationship of organisms that occupy the same territory but show no appreciable competition for its resources because their needs or their sites of activity rarely overlap. In a woodland of the eastern United States, the giant caterpillars of the handsome silkworm moths can satisfy their tremendous appetites without competing for food so long as those of cecropia stick to willows, maple, and shrubs of the rose family, those of luna to walnut, hickory and persimmon, those of polyphemus to basswood, elm, sycamore, and birch or beech, and those of promethea to spicebush, tulip tree, wild cherry and members of the laurel family.

Yet neutralism is possible for animals relying upon the same plant in the same place at the same time. Inherited features, generally in behavior, can still keep their ecological niches separated. A fine example with this effect is known to school children on the main islands of New Zealand. The plant is a coarse native member of the lily family, known as flax (*Phormium tenax*); it is familiar in many other parts of the world to which it has been introduced as a decorative perennial of handsome size. Its narrow leaves, three feet long or more, have a midrib as well as fine lengthwise veins. In New Zealand this foliage attracts two different moths, which lay their eggs on it. The caterpillars of one (a noctuid) eat out only notches from the edge of the leaf, whereas those of the other (a geometrid) cut long narrow "windows" close to the midrib. Although coming within a quarter of an inch of one another, these caterpillars have no occasion to meet. The activities of the one kind interfere in no way with those of the other. Behavioral differences maintain their adaptive neutralism.

In other lands a single tree can support a robin's nest and produce fruit that robins eat without precluding that a woodpecker finds on it insects to eat and a dead part in which to excavate a nest cavity. Birds of the forest maintain their neutralism during the nesting season, when they must all find particularly large numbers of insects as food for their young, by remaining at different elevations above the ground. In eastern North America, blackburnian warblers forage and nest among the treetops, red-eyed

The giant coastal redwood trees of California produce an antibiotic substance in their bark which inhibits the growth of lichens, mosses, and other plants that would be likely to colonize the surface. This example of amensalism is probably related to the ability of a redwood tree to resist an occasional ground fire. With no coating of dry plants on the bark, the fire has no way to spread upward.

vireos in the lower canopy, redstarts and prairie warblers in shrubs of medium height, and hooded warblers on or near the soil. Black-and-white warblers nest on the ground too, but seek insects in bark crevices up and down the trees. Downy woodpeckers dig more deeply into the bark, reaching insects that the warblers miss. Wood pewees dart after insects flying between the trees.

Amensalism

Plants of many species release chemical compounds that inhibit the growth of other organisms, including in some instances the youngest members of the same species. Called amensalism to indicate an unshared table, the relationship reduces the number of competitors for energy and other resources from the environment. Probably the phenomenon evolved long ago, for the chemical compounds (known as antimetabolites or antibiotics) are apparently wastes from normal metabolism. For years, botanists have called these substances "accessory compounds," since they serve no known role in the life of the plant other than to make it distasteful or poisonous to herbivorous animals or to regulate populations of other plants that might offer competition.

The resins in coniferous trees and the milky juice in milkweeds (*Asclepias*) and members of the spurge family (Euphorbiaceae) are effective against many herbivores. The milkweed bug (*Lygaeus*) and milkweed caterpillar (larva of the monarch butterfly) tolerate these compounds and appear to benefit from the repellents by being distasteful to most carnivores.

North American species of spruce and fir, which supply a major part of the pulpwood used in paper making, contain chemical compounds so closely resembling the juvenile hormones of some insects that they reduce the number of attackers. The effective substances regulate development of the insect larvae, preventing them from reaching maturity and reproducing. The existence of this "paper factor" was discovered when, to simplify sanitation of caterpillars being raised in the laboratory, paper towelling and then newsprint (the *New York Times*, and others) were used to cover the cage floors. Strangely, *The Times* (London) and other papers made from pulpwood of Eurasian origin do not have this effect.

Many fungi, particularly those that act as decomposers in soil, produce antibiotic substances of considerable power, effective in restraining the growth of soil bacteria that would otherwise compete. An antibiotic substance in the bark of coastal redwoods in California helps keep these trees inhospitable to clinging plants, including lichens. A different antibiotic substance (chlorellin) is produced by the unicellular green alga (*Chlorella vulgaris*) that holds greatest appeal as a potential food, even indirectly, for mankind; it inhibits the growth of other algal species in the same fresh water.

More complex poisons, released by certain reddish dinoflagellate protozoans, are partly responsible for the epidemic death of fishes, fish-eating birds, sea turtles, squids, and other animals during periodic phenomena called "red tides," when these marine microbes multiply prodigiously.

Predation

Of all the food relationships between unlike species, that between a predator and its prey is the most universally recognized. Often, although not necessarily, the animal

For almost a century, the Hudson Bay Company in Canada bought from trappers and hunters for resale the skins of both the Canada lynx and its principal prey, the snowshoe, or varying, hare. Ecologists have plotted the fluctuating numbers of both animals (see graph, right) from Company records, but cannot tell whether the changes in the number of hares is caused by changes in the number of lynxes or by other factors. Since lynxes sometimes find other prey easily when the hares are few, the curve showing lynx numbers occasionally peaks unpredictably.

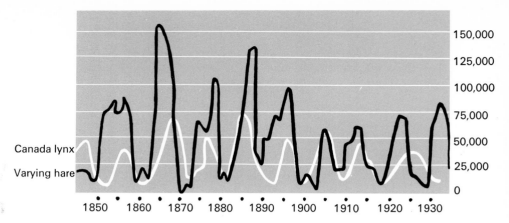

Canada lynx	
Varying hare	

150,000
125,000
100,000
75,000
50,000
25,000
0

1850 1860 1870 1880 1890 1900 1910 1920 1930

that kills and eats another is larger than its victim. The killing seems an essential part of the action. Yet the "birds and beasts of prey" usually kill only when they have a good opportunity and no easier way to get meat. They will scavenge on dead bodies if these are available, or may nourish themselves on plant materials if they can find nothing else. The distinction between a carnivorous and herbivorous animal is an oversimplification; most herbivores eat a few animals and most carnivores a few plants.

Ecologically, a carnivorous plant (such as a bladderwort, a sundew, a Venus'-flytrap, or a pitcher plant) and a herbivorous animal (such as a seed-eater) that kills the plants it eats both fit the specifications of a predator. By contrast, a filter-feeding animal, such as a bivalve mollusk or a baleen (whalebone) whale, is essentially omnivorous, and differs chiefly in being unaware of the individual living things it eats. Filter-feeders respond to quantitative and qualitative changes in the food they get, rather than to the species or individuals by any criterion other than size and mobility.

In the realm of predation, the opposite extreme to herbivorous habits would be cannibalistic ones. Cannibalism is actually more widespread than was once believed, and often serves importantly as a regulator of population growth. Under natural conditions, as well as in tropical aquariums, the guppy fish (*Lebistes*) often devours its own young. Large bass eat small bass of their own species. Crocodiles become cannibals when other food is scarce. Many birds, under conditions of stress when populations are large, destroy their own eggs and eat the contents. In crowded quarters, similar stresses induce rats and mice to kill and eat the helpless young, even when other food is abundant.

Field studies show that the predators choose the easiest prey available. Generally these are the young, the sick, the victims of accident, the aging, and the nonconformists for whatever reason. Inevitably the predators slow the rise and round the peak of numbers in the prey population. Seldom, however, can they hold the numbers of the prey to the practical limits imposed by the productivity of the green plants in the area. The availability of prey regulates the population of predators more often than the number of predators limits the abundance of prey.

Prey animals are often well adapted in relation to their commonest predators, which excel in acuteness of senses, live longer, learn by experience, and are highly selective among potential victims. Members of prey species benefit from protective coloration, which may be as simple as countershading or as amazing as camouflage and simulation of inedible features in the environment. Prey animals excel in remaining immobile, inconspicuous, and apparently odorless. Others are actually repugnant in some way, and marked conspicuously or deliberate in movement as though warning predators to avoid them. Because predators learn to do so, imitators that have warning coloration and deliberate movements (but no venom, malodorous or toxic substances to give real protection) can benefit through mimicry of the type that Henry Walter Bates discovered among insects of the Amazon basin.

On the high plains and savannas of equatorial Africa, the antelopes of various kinds and the zebras tend to travel and feed in herds. Each group of prey animals benefits from the combined awareness of its members, and takes to its heels if hungry predators such as lions are detected. Often

So long as the source of nutrition is the body of an animal, a carnivorous organism may be a predator that kills its prey, a scavenger that finds dead prey, a parasite, or even a plant. A warthog in Rhodesia becomes the prey of lions (top, right) but is not big enough to satisfy several predators for long. After feeding on larger prey, such as a giraffe, the lions wander off, leaving large amounts of meat to be scavenged by maribou storks (above) and other opportunists. In southern Europe, the griffin vulture soars above wilder mountain country until it sees a carcass, such as that of a deer, then descends to make a meal (top, left). Entirely different behavior serves the red fox of North America (right) as it stalks a rodent, such as an unwary ground squirrel. Carnivorous plants live mostly where nitrogenous nutrients are scarce and use their roots for anchorage while capturing and digesting the bodies of small animals for nitrogenous compounds. The flexible leaves of a sundew (far right) slowly curl around a captured insect and secrete digestive enzymes against its surface. When absorption is complete, the leaf uncurls and drops the remains.

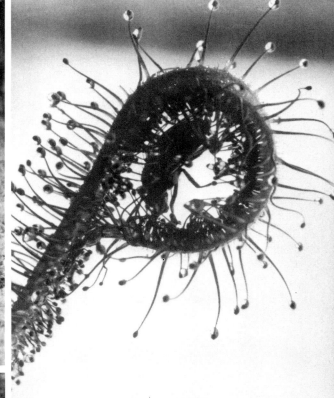

the brindled gnus (blue wildebeests) and zebras maintain loose associations in which the superior vision and poorer olfactory sense of the zebras is counterbalanced by the superior olfaction and lesser visual abilities of the gnus. Yet the greatest advantage in all large groups may lie in the fact that individuals at its core are relatively protected. The individuals in peripheral positions are the ones most exposed to predators. This is true also of fishes that swim in great schools and "armies" of fiddler crabs feeding in close order on a mudflat.

The size of a feeding group is limited by the availability of suitable food. Only if nourishment is abundant can numbers be large. With an increase in the population or a seasonal decrease in the resources, each individual must travel farther to get enough to eat. The whole group may travel faster, letting infirm individuals, and perhaps young too, lag behind and be pounced upon. Or the oversized group divides, and the fragments progress at a more leisurely pace.

The fate of herds that shrank too far was often observed on the plains of North America while bison and pronghorns were numerous. When wolves harassed a group of a dozen or fewer adults, they seemed unable to form a compact circle, facing outward toward the predators. The young had no place to cower at the center of the group. Instead, the herd tended to panic and stampede. The wolves and coyotes then picked off the immature individuals, next the slowest adults, and finally the last of the little group.

The presence of prey in large numbers for only a limited time confers a significant advantage in safety from predators. When the countless waterfowl and shore birds arrive to nest on the arctic tundras, the arctic foxes and other predatory animals feast, then stand around or lie quietly until they digest their meals. Food is far more abundant than they can use. Yet the number of predators cannot increase significantly since it is limited by the scarcity of food in winter when the migratory birds have departed. Recently the same advantage was recognized in the sudden emergence of 13-year and 17-year cicadas in concert, after their long development in the soil. In the intervening years since the parents of these broods shrilled loudly, mated, and laid their eggs, the predators that had feasted on the cicadas had all died or moved elsewhere. As soon as the insectivores are sated, the surviving cicadas can be as noisy and conspicuous as is helpful in finding mates without inviting additional attack.

Parasitism

Probably no one will ever know how far back in time the first parasite began attaching itself to an organism of larger size and taking nourishment from it without waiting for death to occur. The essential feature of a parasite is its relative size— smaller than its host—with the corollary that, if well adapted, it can gain its ends without sickening its host, let alone killing its source of food. Disease symptoms in the host may indicate that the parasitic relationship is relatively new, and the parasite poorly fitted to its ecological niche.

Ecologists generally assume that the parasitic habit has been derived from other types of interaction between members of unlike species. Probably the parasitic bacteria and parasitic fungi originated as decomposers. Many of them maintain small populations on or in the bodies of potential hosts for years, without being able to fully express their biotic potential unless the host weakens from malnutrition. Whether the ultramicroscopic viruses, of which all that have been discovered are parasites, evolved by degeneration of cellular organisms (such as small bacteria) is far from clear. Since evolution is still in progress, decomposers must still be adding ways to shift from a dependence upon dead cells to a utilization of organic compounds from organisms that are weakening toward death and unable to repel the decomposer. A further shift, allowing successful attacks on healthy organisms, would be possible for some, establishing a new parasitic relationship.

A scavenger might upgrade its source of nourishment from animal products and dead bodies to living organisms, becoming a parasite rather than a predator or a herbivore. The flatworms (phylum Platyhelminthes) can be arranged in a logical series suggesting the sequence in adaptive changes needed to shift from scavenging to obligatory parasitism. Equally striking are the ways in which the inherited behavior patterns of parasites match minutely the behavior of the hosts used in each life history, and the physiological adaptations that so often restrict the parasite not only to one type of host but to a single organ or a small area for attachment. These features can be appreciated in the life histories of tapeworms, roundworms, thorny-headed worms and other internal parasites. Their specificity in hosts is challenging. Each order of birds, for example, has its own species of tapeworms, as though the parasitic relationship had evolved before the diversification of birds. More probably the limiting feature is in biochemical similarities among the related birds, matching the tolerances of the parasites.

Even parental care may have been an avenue to parasitism. This route is suspected among insects of the order Hymenoptera. A great many kinds merely prey upon other insects and spiders, preparing a store of still-living victims for their own larvae. It seems only a step beyond this to the habits of ichneumonflies, which lay their eggs on caterpillars and other hosts within which their larvae will feed, generally not killing the host until ready to pupate.

Insects that parasitize the eggs and young of other insects are called parasitoids. Most are highly specific in their choice of hosts. This has enabled economic entomologists to propagate particular kinds of parasitoids—chiefly ichneumonflies (family Ichneumonidae) and chalcids (Chalcidae) and tachina flies (Tachinidae, in order Diptera)—to promote the biological control of pest insects.

Animals that have evolved the habit of imbibing blood or plant juices become potential agents (vectors) for transferring still other parasites that can ride along from one host to the next. Apparently an aphid (plant louse) acts as no more than an animated hypodermic needle as it transfers mosaic virus from a diseased tobacco plant to a healthy, susceptible one. Some vectors, however, serve as reservoirs—even generation after generation—for microbes that cause disease when they reach a suitable host.

The evolutionary origin of the complex ecological relationship between a parasite and its vector is often mysterious. Is the malaria organism, for example, the descendant of parasites that formerly reached vertebrate hosts (reptiles, birds or mammals) in contaminated water, went through sexual stages in the intestine, and then invaded the blood stream? Perhaps the

The borderline between the epiphytic habit of orchids that grow on trees but absorb no nourishment from them (left) and parasitism tends to vanish if the epiphytes intercept the solar energy the tree would otherwise use. Many leeches become blood-sucking parasites (right) on frogs but survive, between blood meals, as predators on small arthropods, mollusks, and worms.

descendants found a way to reach the vertebrate blood stream more directly by going through their sexual stages in the gut of a bloodsucking mosquito instead. Or is the malaria organism fundamentally a parasite of adult mosquitoes, using a developmental stage in the blood of a terrestrial vertebrate merely as a step toward infecting more mosquitoes? Almost equally convincing evidence can be cited for either view. Only a long-lived insect can serve as the vector for a disease of this kind. Only such an insect is a suitable host for the protozoan parasite.

No organism seems immune to well-adapted parasites. Parasites upon parasites ("hyperparasites") have been famous ever since 1733 when Jonathan Swift observed in verse that "a flea / Hath smaller fleas that on him prey; / And these have smaller still to bite 'em." (Similarly, both parasites and commensal organisms have their commensals.)

Parasites are well known among the flowering plants, each with its own way to get a continuing supply of nourishment at the expense of some other species. The so-called strangler trees (chiefly figs, *Ficus*) begin as perching plants in warm moist forests. They extend a meshwork of woody roots down the trunk of the big tree that supports them and eventually stand independently on this hollow support after shading the host tree to death and taking its place in the forest canopy.

Most spectacular of the parasitic flowering plants are those that lack leaves and chlorophyll and that display only flowers or flowering stalks above ground. The giant is *Rafflesia arnoldi* of Malayan rain forests, whose roots appear to take nourishment from those of certain climbing vines (*Cissus*); the huge blossom of the parasite

is single, short stemmed, and as much as four feet in diameter. Far smaller and more frequently encountered are the North American snow plant (*Sarcodes*) in western mountains, broomrape (*Orobanche*—named for its relationship with broom), beechdrops (*Epifagus*) in beech woods, and Indian pipe (*Monotropa*) under conifers in mixed temperate forests. Recently, close study has revealed that at least some of them are parasitic upon fungus strands that form mantles around their roots, isolating them from the soil. The fungus may be a decomposer (a saprophyte), digesting the humus components underground. Or, in the case of Indian pipe, the fungus may be getting its organic compounds from the roots of a hemlock or other living conifer; this makes the Indian pipe indirectly a parasite of the conifer, without actually coming into contact with it.

Commensalism

A commensal is an uninvited guest, one that does no measurable harm while gaining a little from the association. Most plants with this habit perch on the high branches or cling to the bark of trees, and thereby live where they can use their own chlorophyll and get what moisture and mineral nutrients they need from rain and dust. A perching plant of this type is called an epiphyte. It may be an orchid, a bromeliad, a fern or liverwort, a moss or a lichen.

Commensal animals seem often to gain merely a protected living space, such as the vacant portion of the burrow made by a marine worm or of the mantle cavity of a bivalved mollusk. Specialized crustaceans slip in and out of these spaces. The pearl fish (*Carapus*) slides backward into the respiratory openings of sea cucumbers (holothurian echinoderms), bivalves or tunicates, often limiting itself to a single species and size of host.

Over much of the American West, burrowing owls of medium size used to be frequent commensals in prairie dog towns. They took shelter when the sun was high and even nested in the underground passageways excavated by their rodent hosts, without interfering noticeably with any activities. Now that ranchers have so largely exterminated the prairie dogs, in the belief that this action would leave for domestic cattle the vegetation on which the rodents formerly fattened, the owls have

become scarce. This change reduced the
predation by owls on grasshoppers, other
insects and mice—herbivores that now are
far more numerous and destructive of the
vegetation.

Some commensals ride about on their
hosts, reaching fresh supplies of food with
a minimum of effort. The animals that are
carried are epizoans, and their habit is
called phoresy. Some attach themselves
temporarily, as does the pilot fish (or
remora or shark-sucker, *Echeneis*) that
uses a highly modified dorsal fin to cling to
the body of a large shark or sea turtle; it
frees itself and feeds on scraps when its
host stops to eat. Certain sea anemones
live as epizoans on the snail shells in which
active hermit crabs protect their soft abdomens while moving about. Barnacles
that grow on the skin of whales seem
attached permanently, and often become
parasites by absorbing nourishment from
the host that carries them.

The boundary between commensalism
and either parasitism or protocooperation
is often ill-defined. Small beetles live in the
nests of ants and of termites, acting and
perhaps smelling enough like their colonial
hosts to be tolerated. The beetles solicit
food from their hosts, or help themselves
to the contents of storage chambers. In
seasons when the colony is well supplied

with food, it is easy to regard the beetles as commensals. In seasons when the hosts need the food for growth and reproduction, the beetles are parasites. But in many instances, the beetles approach their hosts and regurgitate droplets that the ants or termites accept promptly. In exchange, the host may then offer the beetle regurgitated food, making the relationship one in which energy is transferred in both directions.

Protocooperation

The distinguishing feature of proto-cooperation is that, although both of the species gain extra energy or survive longer when together, each can live apart from the other. The common association be-tween brindled gnus and zebras fulfils these requirements, since the combined herd can take more effective evasive action from predators than when the species remain separate.

On the same African plains, the oxpecker birds (*Buphagus*) perch on antelopes of many kinds, on giraffes, rhinoceroses, and elephants. They run over the hide, finding ticks and sometimes leeches as food. Buff-backed herons (*Bubulcus iris*, known in America as cattle egrets) perch in the same way, but are less persistent in hunting for parasites. All of these birds, however, maintain from their vantage points an alert watch for large predators and people approaching from any direction. Through actions and sounds, the birds warn the mammals, often in time to permit escape or in ways that direct a counterattack.

In coastal seas, some of the crustaceans have evolved the adaptive habit of shielding or camouflaging themselves with living organisms of other kinds, including small sea anemones, tufts of colonial hydroids, small tube-dwelling worms, and bits of sponge. The crab may hold these normally-attached kinds of life on its own back until they take hold. Experiments have satisfied ecologists that the crabs obtain in this way a statistically significant decrease in attacks by predatory fishes and octopuses. Probably the organisms the crab carries gain also, in opportunities for feeding.

The concept of protocooperation can be applied to plants too. Examples are numer-ous in which flowering plants display colorful parts and offer odorous materials, nectar and expendable pulpy tissues that attract pollinators and fruit-eating animals. These rewards for visitors increase the efficiency of pollination and of seed dis-persal. Usually the relationship is non-

Small prairie owls (bottom) take shelter, when they can, in the burrows of prairie dogs (top). Since the owls emerge to find mice and insects as food, they do no harm to the prairie dogs and are regarded as commen-

sals. Prairie rattlesnakes that take shelter in the same burrows may occasionally prey upon prairie dogs, although the snakes are active at night and the prairie dogs only by day.

specific, in that quite a number of different animals visit, and the same animals visit different plants in season. Both the plant and the animal can get along without the visit, the one reproducing by vegetative means and the other finding alternative foods.

Mutualism

The distinguishing feature of a mutualistic association is that at least one partner is unable to grow normally or reproduce without the other. To this extent it is an obligatory relationship. The partners may be unlike plants, such as the fungus and the alga that form a lichen, or the bacteria that live in the root nodules on leguminous plants, or the fungi that surround the root tips of a pine tree. The association, instead, may involve a plant and an animal, such as the algae and coral animals that together form reefs, or the rumen bacteria that enable a cud-chewing mammal to get nourishment from cellulose, or the *Pronuba* moth that alone can pollinate a yucca plant in the natural way. Or the partners may be unlike animals, such as the protozoans that digest the cellulose a termite swallows, or the decoy fish that lure victims to the big sea anemones in the South Pacific.

The oldest and most widely distributed of mutualistic relationships between plants is almost certainly that in lichens. Each lichen consists entirely of strands of fungus, which hold the partnership in place, capture moisture and protect it under a skinlike layer, and of algae as separate cells, which use the water in photosynthesis, elaborating organic compounds needed for the growth of both partners. Some lichens tolerate salt spray and grow on rocks close to the level of the highest tides. Many inhabit the arctic tundra, or grow near the permanent ice fields on the tallest mountains. Some live on Antarctica, where intense cold is normal. Still others cling to poor soil, tree bark, stone walls, and tombstones in most parts of the world. A great number of lichens grow as perching plants in rain forests of the tropics and the temperate zones. Probably lichens were the earliest form of vegetation to colonize the continents and islands of the world.

So characteristic is the growth form attained by each pair of species—fungus and alga—that the binomial system of naming plants has been extended to these partnerships. Reindeer "moss" is *Cladonia rangiferina*, smooth rock tripe *Umbilicaria mammulata*, and map lichen *Rhizocarpon geographicum*. Only a lichen specialist is likely to know the true identity—genus and species—of each of the partners. Only a few are club fungi, the rest being sac fungi. Most of the algae are single-celled greens (phylum Chlorophyta); some are blue-greens (Cyanophyta). In suitable situations, which rarely are out of fresh water, the algae live apart from the fungus and grow normally. None of the fungi live apart, although they can be cultured on special nutritive materials. Some biologists regard this as proof that the fungi are parasites and the algae the hosts. Others point out that neither organism alone can withstand the extreme and prolonged desiccation that many lichens endure routinely.

An equally intimate relationship has evolved between microbial plants and vascular ones, in which specific bacteria or actinomycete fungi invade the roots or leaves and there contribute to the nutrition of the conspicuous plant. The microbes capture atmospheric nitrogen and incorporate it into nitrogenous compounds such as amino acids, which are shared with the

Protocooperation

Two kinds of starling-like oxpeckers associated with large native mammals and with domestic horses and cattle on the high plains and savannas of equatorial Africa. These birds clamber all over and under their large associate, devouring whatever blood-sucking ticks and flies they find. The large mammals not only tolerate these attentions but heed the movements of the birds, and particularly any sudden departure, using the oxpeckers as lookouts. The black rhinoceros (left) can thus compensate for its own poor vision, and the African elephant (right) can learn of possible danger coming from behind, where its own bulk interferes with its awareness.

Often the presence of small animals on the bodies of large ones indicates a parasitic relationship. The rice-sized objects attached to the body of the sphingid caterpillar (left, above) are the cocoons of a parasitic wasp (Apanteles) the larvae of which have nourished themselves inside their host. The wrasse (top) is being cleaned of external parasites by large marine crustaceans —a relationship benefiting both. Large herbivorous mammals are similarly tolerant of birds that pick off ticks and also serve as sentinels: a buff-backed heron on a Cape buffalo (left, below) and several starling-like oxpeckers on a vigorous bull of the sable antelope (immediately above).

105

Each lichen partnership, composed of algal cells enmeshed in fungal strands, has a style of growth and pattern of tolerances that fit it for a particular ecological niche. The boulder lichen (Parmelia) *has a lobed, leaflike appearance as it spreads over rocks or expands on the bark of trees and wood that has been exposed for many years (below, far left). The pyxie cup or goblet lichen* (Cladonia) *grows erect to a height of an inch or less on poor soil as well as on rotting wood and rocks (below, left). Old-man's-beard lichen* (Usnea) *forms pendant branching streamers below tree branches in wet woodlands (below, right). The partnership between the* Pronuba *moth and the* Yucca *flower (far right) is more obvious because the unlike individuals that associate with mutual benefit are big enough to be easily observed.*

vascular plant. In exchange the microbes receive carbohydrates, water, and probably hormones. Usually the clusters of these nitrogen-fixing microbes in the root tissues grow into gall-like nodules, such as are found on the roots of most legumes—members of the second-largest family (Leguminosae) of flowering plants. Nodules of similar form and probable significance occur also on the roots of alders, cycads, and many other woody plants.

Strands of different fungi are essential for the normal growth of the most primitive of vascular plants (the psilopsids), the coniferous trees, the terrestrial orchids, most members of the heath family (Ericaceae), and many others. The fungi, which are either alga-like fungi or club fungi, form characteristic mantles around the root tips, actually isolating them from the soil. While the fungus attends to the absorption of essential mineral substances

for the partnership, it draws organic food from the conspicuous plant. Many of these fungi, which include such well-known mushrooms as *Amanita*, *Boletus*, *Lactarius*, and *Russula*, seem unable to form their own reproductive bodies unless associated with the roots of specific living trees.

The need for these fungus organisms, called mycorrhizae, became evident recently in Puerto Rico, when young pine trees set out in an afforestation program failed to grow properly. A teaspoonful of soil from a northern pine forest was sprinkled on a test area, inoculating it with the appropriate fungi. In this area the pine trees began growing at a rate of nine feet annually, whereas the other pines around them continued stunted until the fungi spread and established the partnership.

Coral animals of the types that produce fringing reefs in warm seas are even more dependent upon certain algae. As was discovered a few years ago in Jamaica, the algal partners must be healthy and active in photosynthesis if the coral animals are to secrete their characteristic limy external skeletons. This restricts formation of coral reefs to sunlit depths. The complex chemistry of the interaction is not yet completely understood. The photosynthetic activity may contribute essential organic compounds. Utilization of carbon dioxide may affect calcium metabolism. The algal partners are known as zoochlorellae if green, and zooxanthellae if their chlorophyll is masked by supplementary brown or yellow pigments. The algae themselves prove to be unicellular greens or dinoflagellates identical with those that live and multiply in sea water apart from the corals. Similar algae are maintained as partners within some other coelenterates, sponges, marine flatworms and mollusks.

On land, the role of microbial plants in the nutrition of vertebrate animals is becoming increasingly well known. At first, the rumen bacteria that live in the rumen (paunch) of cud-chewing mammals (ruminants) seemed unique. They split cellulose into small, digestible, and absorbable molecules, allowing the mammals to get energy that otherwise would be unavailable from this abundant chemical component of their herbivorous diet. Without their rumen bacteria, the cud-chewers could not survive the dry season on the grasslands of the world, where they eat dead grasses and other undecomposed remains of vegetation.

Probably a majority of animals make use of substances released by bacteria in their intestinal flora. Mutualistic microbes in the human small intestine synthesize vitamin B_{12}, but only when they receive an unidentified "intrinsic factor" secreted ordinarily by the host. An inherited inability to secrete this factor, or a lack of the appropriate bacteria (perhaps following antibacterial treatments with antibiotic drugs) leads to a form of malnutrition. It becomes evident because the afflicted person is unable to synthesize essential components of the blood-clotting mechanism.

Flowering plants whose dispersal is dependent upon producing seeds include a number that have become mutualistically related to particular pollinators. Only female moths of one genus (*Pronuba*) gather the sticky pollen of yucca flowers in southern North America, and transfer it. This adaptive behavior of the moth includes laying a few eggs in each pollinated flower. The caterpillars that hatch out use as food a small number of developing seeds in the yucca fruit—seeds that would not have developed without the moth. Some orchids attract only the males of specific wasps, and stimulate the insects to perform mating behavior in ways that cause them to pick and transfer pollen; supposedly the insects also gain from this activity.

Of the mutualistic relationships between unlike animals, the most widely distributed may be that between single-celled protozoans of the flagellate order Hypermastigida and certain termites and wood roaches. These flagellates include some with the unusual ability to simplify molecules of cellulose and to gain energy from this activity. In the intestine of wood-eating insects, the flagellates digest the wood fibers that the insects chew up and swallow. Neither partner can live alone.

Not all of the hypermastigid flagellates present in the gut of a termite have this ability to be mutualistic partners. As L. R. Cleveland at Harvard University learned from experiments with the termites of New

England, three of the seven kinds of protozoans are commensals that contribute no nourishment to the host inside which they propagate. At least one of the other four kinds must be present to keep the termite from starving.

A termite can be "defaunated" by heating it to temperatures that are lethal to the flagellates but not to the insect. It will replenish its intestinal fauna if it can associate for a short time with an untreated live termite, or even walk about where untreated termites have been recently. Insects of this kind keep themselves well supplied with their intestinal partners by exchanging droplets by regurgitation and by frequently eating fresh fecal pellets; they get live flagellates in both ways. One reason for these actions is that a termite loses all of the contents of its digestive system at every molt, and must replace its essential associates. Much of the social organization of a termite colony seems to serve in assuring a continuation of the mutualistic relationship.

Among marine life, mutualistic associations have evolved between certain fishes and coelenterates. One of the most amazing has been studied in coral reefs of the Indo-Pacific region, where a damselfish (*Amphiprion percula*) earns a reputation as a decoy, inviting predatory fishes to their death among the tentacles of a big sea anemone (*Stoichactis*). Actually, the anemone will not spread its tentacles fully unless brushed against by the damselfish, which is unharmed by the stinging cells. Larger fishes pursue the decoy, which is brilliantly striped, brown and electric-blue, only to bump into the outstretched tentacles and be caught. Then the damselfish can feed on the immobilized prey while the anemone works the victim into its mouth.

Interrelationships

One goal of ecologists is to be able to make reliable predictions about organisms living under natural conditions. Most often they are asked to forecast the fate during the next decade or two of some animal that seems on its way to extinction, or of some new colonist that has just begun to make a place for itself. The need for full information is no less today than in 1859 when Darwin concluded that:

> Battle within battle must ever be recurring with varying success; and yet in the long run the forces are so nicely balanced, that the face of nature remains uniform for long periods of time, though assuredly the merest trifle would often give the victory to one organic being over another. Nevertheless, so profound is our ignorance, and so high our presumption, that we marvel when we hear of the extinction of an organic being; and as we do not see the cause, we invoke cataclysms to desolate the world, or invent laws on the duration of the forms of life!

The sole surviving species in the reptilian order Rhychocephalia merits special consideration. Known by its Polynesian (Maori) name as the tuatara (*Sphenodon punctatus*), it inhabits steep volcanic islets around the coast of New Zealand, where it takes shelter in nest burrows dug by shearwaters (birds of genus *Puffinus*) among the roots of native evergreen shrubs (*Coprosma* species, of the madder family, Rubiaceae). Each year the burrow-making activities of the shearwaters loosen the soil and work their droppings into it, providing fertilizer where it stimulates growth of the *Coprosma*. The foliage breaks the force of frequent storms and shades the soil, preventing rain and sun from making it too hard for the birds to burrow into. These relationships might be indicated as

Coprosma - Puffinus	+	+	Protocooperation
Coprosma - Sphenodon	0	0	Neutralism
Puffinus - Sphenodon	0	+	Commensalism

If domestic goats are introduced on these islands, as they have been on some, they eat the *Coprosma* foliage and tender twigs so intensively that the shrubs die. The soil hardens. The shearwaters cannot dig burrows and they fly elsewhere. The tuataras have no shelter and fail to survive. The land snails and giant wingless crickets on which the tuataras ordinarily feed at night then multiply. Yet the associations shown by the goats are neutral, except for their herbivorous predation. This one addition soon alters the whole ecosystem on the island.

The probable relationship of a newcomer among populations of established species is equally important. So often the animal or plant arrives beyond its native range through human action, whether deliberate or not. The new arrival may fail to maintain its beachhead by finding an ecological niche in which to compete successfully. It may, instead, spread as European house sparrows and starlings have since their introductions in 1850 and 1908, mostly by replacing native birds. In this instance, the principal populations affected were those of bluebirds, tree swallows and the smaller woodpeckers, all of which proved to be less vigorous than the introduced birds in competing for nest holes.

A newly arrived green plant may compete successfully with plants man values more, and become a weed. Or the colonist may have a disease to share. Mosquitoes that reproduced in the barrels of fresh water on the ships of slave traders coming to America from Africa brought yellow fever to the New World, where the infection quickly found new hosts in arboreal monkeys and in man. Infections from wild animals that spread to mankind or domestic livestock are called zoonoses (each a zoonosis) and provide topics for intensive study.

Two conclusions can be reached from ecological research on interspecies relationships. The newcomer to fear is the one that is prolific, eats almost anything, and needs a minimum of shelter. To resist such a colonist, a community must be rich in species and thereby strong with alternative pathways for the transfer of energy and the cycling of materials. The avenues of interaction that make an ecosystem so intricate confer upon it the stability that gives it a future. As the distinguished American biologist Barry Commoner wrote, "The web of relationships that ties animal to plant, prey to predator, parasite to host, and all to the air, water, and soil which they inhabit persists *because* it is complex."

8 Growth of populations

Under most circumstances, the world has no room for the extra offspring. In one way or another, the surplus is eliminated, generally benefitting species other than the one that reproduced so extravagantly. Until ecologists recognized that the unsteady balance of nature is sustained by a regular inflow of solar energy and the cycling of chemical substances through a multiplicity of lives, the economy of the biosphere appeared as improbable as that of the town whose citizens all claimed to make a living by taking in one another's laundry.

Success for a species depends upon the survival of so many offspring that the parental rate of reproduction is maintained at least for another generation. For new individuals to make their appearance is not enough. The size of each population depends not only upon natality but upon the loss of individuals by dispersal (emigration), the arrival of individuals from elsewhere (immigration), and the failure of individuals to survive (mortality). Mortality, like the balance between emigration and immigration, is a measure of the resistance shown by the environment toward an increase in population. For the numbers of a species to remain fairly constant, fluctuating above and below an average total, the environmental resistance must approximately equal the biotic potential.

The factors that regulate the size of a population are so numerous and varied that ecologists feel a need to study them separately before trying to understand them in combination. Some are nonliving aspects of the environment, such as adverse weather, unsuitable soil, and the prevalence of poisonous materials. Others are clearly due to living things of the same and different kinds: unsuitable food, predators, parasites, and debilitating or deadly diseases.

Natality

For relatively few kinds of plants and animals are the vital facts about potential reproductive rate known with any certainty. One of the most familiar is the English vole (*Microtus agrestis*), often called a meadow mouse. Of the 200 to 300 that live on the average acre of farm land, about half are females. At 30 days of age a female begins breeding, a male at 45 days. Her litters of 4 to 8 young are born after a gestation period of 21 days. Ordinarily she is pregnant again within minutes after giving birth. At age 6 months she reaches her peak of fertility and, at best, survives for 18 months. By then her reproductive abilities end through senescence. During her lifetime a female may have 24 consecutive pregnancies and give birth to 150 young. While biologists generally assume that any species will reproduce as fast as is possible, the full potential is rarely realized. Recently a California species of the same genus and general characteristics did increase to a population of 12,000 per acre—a pair for every 7 square feet. The voles destroyed the vegetation, overturned the topsoil, and died with no known survivors.

The number of eggs or of young is sometimes cited as a measure in comparing the reproductive possibilities of various animals. The elephant may have one offspring every other year, a woman from 1 to 5 surviving children from a single birth, a nine-banded armadillo 4 identical quadruplets each year, an alligator 10 to 25 eggs, a lobster 12,000 to 15,000, a sturgeon 3 million (together weighing almost a third as much as the parent), an oyster 100 million to 500 million. Yet all of this effort leads in the average lifetime of a parent to only as many survivors as are required to take the parents' places in perpetuating the species.

The common American crow, which may choose one mate for life and cooperate in nest-building and raising a new family each year, spends considerable time sociably and noisily communicating with the other bird. Animal behaviorists speak of such activities as "reinforcing the pair bond." Ecologists see it as a step that contributes toward the welfare of the young and the successful perpetuation of the species.

For any kind of life that reproduces sexually, the number of generations in a century has more importance than shows in the census figures for the population. Each generation is a new opportunity to sort out and recombine the genetic heritage, increasing the variability of the species and testing possible combinations against the tolerances required for survival. The fruitfly (*Drosophila melanogaster*) studied by geneticists produces about 200 offspring in 11 days from each mated female. In the years since these flies took on scientific importance, they have gone through more than 2,000 generations. They have had more chance to evolve during the twentieth century than man has had since 1400 A.D. The housefly (*Musca domestica*) may go through annually only 10 generations, instead of 33. Yet the houseflies of the world needed less than 50 generations to evolve a massive resistance to DDT, then tolerance for almost any insecticide that was cheap and reasonably safe to use.

Mortality

For any population that is utilizing its resources fully, the age at which each individual dies holds immense importance. The death of an old one, who has already contributed fully to the future of the species, is least significant. Elimination of a very young individual deprives the world of any possible progeny. Yet each of these deaths improves the competitive position of all of the survivors as they vie for space and energy.

Under natural conditions, three different patterns of mortality can be recognized. Our own appears to be an extreme development of the pattern that has been traced carefully in the Dall mountain sheep, in

Each species has its own mortality characteristics. The survivorship curve for mankind (I) shows the lowest mortality. That for Dall mountain sheep (II) shows that less than 7,000 out of each 10,000 born reach reproductive age. In a few animals, such as freshwater hydras, the number of survivors decreases in proportion to the number present (III). By far the most common pattern involves high initial mortality (IV), followed by a reproductive period with fewer losses.

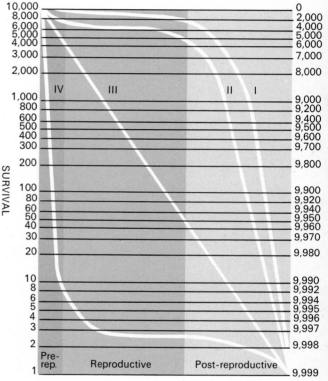

LIFE SPAN

which a burst of deaths soon after birth is followed by improved survival into a post-reproductive range of ages. These individuals seem to have a fair chance to live out a normal lifespan before the death rate increases sharply again.

With protection from ordinary hazards, sheep of this kind in a zoo commonly live to an age of 15 years. In Mount McKinley National Park, where Dall sheep are native, the average age at death was 7 years among 608 whose skulls Adolf Murie picked up and tallied from the annual rings that marked the curving horns. The death rate is highest among the lambs, which have difficulty in following the full-grown animals across the mountain slopes, and

The nine-banded armadillo of northern Latin America and the southern United States regularly bears identical quadruplets. Human reproduction generally provides a single child at a birth from a single egg (right), with which a single sperm has fused. Zoologists are often impressed by the considerable distance, compared to its own length, that a human sperm will swim against the current (far right) to reach an egg in the Fallopian tube.

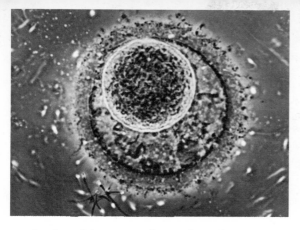
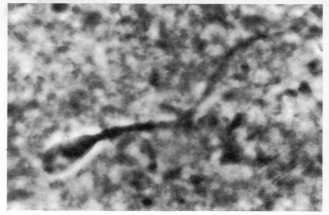

again in old age, when the sheep grow weak from malnutrition because their teeth are worn and broken. At intermediate ages, the death rate diminishes because the sheep escape more easily from the wolves that are their principal predators.

The mortality rate for mankind in countries with modern medical care remains low from the first year after birth to well beyond the reproductive period. Only then, and for a complex of reasons, it increases again, bringing the survival curve down much more sharply than that of the sheep. Customarily these changes in the death rate at different ages are made evident most easily on graphs showing both survival and mortality on a logarithmic scale in relation to lifespan.

For living things as unlike as American robins and the freshwater hydras, the death rate is relatively constant throughout life. It is as though accidents were the chief causes of mortality, striking equally at all ages. Less than a third of the individuals may reach reproductive maturity.

By far the commonest pattern of survival is that in which the mortality begins high, and then decreases until later. Less than a twentieth of one per cent may survive to reproduce. This is the pattern in oysters, most fishes, and many insects. On this schedule, as in the pattern where the death rate is almost constant, the average age of an individual bears almost no relevance to the age that could be regarded as a full lifespan. The average age shows mostly the dimensions of the delay, whether hours or days or weeks, before half of the population is dead, making energy and nutrient materials available to other living things. Often ecologists refer to this delay as the "turnover time" of the species.

Many attempts have been made to find a formula relating average mortality rates to the average lifespan. None of the suggestions made so far have any rational basis. Yet George A. Sacher at the Argonne National Laboratory in Illinois did find in 1959 a way to account for the fact that small horses, small dogs, and small women live longer on the average than large horses, large dogs, and large men. After many trials, he discovered that the most satisfactory predictions come from a double correlation, with both brain weight and body weight of the adult. This could indicate that the larger the brain is in relation to the body weight, the longer the individual is likely to live. Ponies do have a larger brain in comparison to body size than do large horses, and sometimes live to be 40 years old; 34 years is extraordinary for a horse, and the average is about 22. Similarly, between birth and age 18, a girl's body grows only 4.2 times as fast as her brain, whereas that of a boy increases 4.6 times as fast as his brain. She finishes a head shorter, pounds lighter, but with brain and longevity relatively larger.

Sacher's formula seems to account for all but about 16 per cent of the variability in life spans that can be deduced from the most reliable records for people, other primates, and a few representatives each among rodents, insectivores, carnivores, ungulates, and elephants. Specifically, it relates the logarithm of the average lifespan in years (x), the logarithm of the average adult body weight in grams (y), and the logarithm of the average brain weight (z), in the equation $x = 1.035 - 0.335y + 0.636z$. Thinking of longevity as a desirable goal, one British scientist commented that it is "an advantage to have a brain, but a disadvantage to have a body!"

It matters little whether death is due to a

Differences in reproductive habits and public health lead to age pyramids of diverse shapes. In the United Kingdom in 1959, less than 24 per cent of the population consisted of dependent children under the age of 15, and a slightly larger proportion were oldsters, past reproductive age but to a major degree self-supporting.

In India in 1951, by contrast, the under-15 group represented 37.5 per cent of the population, and the post-reproductive individuals only 12.4. Britain has tended toward smaller families and an average age of 35, whereas India has produced more children but a population whose average age was only 20.

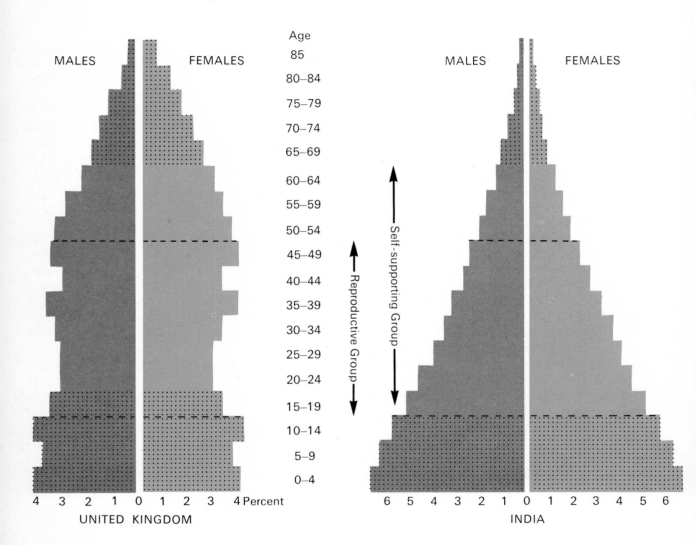

failure of internal mechanisms, or a diminished resistance to disease and predators. It can still be called for as a programmed step in normal development, by a "genetic countdown," as Carroll Williams has called it in insects. Natural selection would presumably favor any line of inheritance that regularly eliminated post-reproductive individuals unless their presence contributed to the survival of younger age groups.

A clear contrast in the inheritance of shorter and longer lifespans was discovered recently by A. D. Blest of the University College in London, among some related species of moths that fly at the same season along the edge of the rain forest in Panama. Those found to be both palatable to predators and colored in a pattern that give them camouflage became hyperexcitable at the end of their reproductive period; they wore off the scales that provided their concealment and made themselves conspicuous, thereby hastening their own death. Species that were unpalatable and marked in patterns that a predator could learn to recognize (and avoid) lived as post-reproductive individuals for 45 days on the average, instead of 13 in the other species; they continued to behave in a normal fashion, and afforded predators an opportunity to learn to avoid them without any risk to the reproduction of the species.

One of the first British zoologists to study factors limiting the human lifespan, G. P. Bidder, suggested that "no man ever reached 60 years of age until language attained such importance in the equipment of the species that long experience became valuable in men who could neither fight nor hunt." Postponement of death becomes worthwhile if the advice of older people can save the lives of younger ones.

The curtailment of human life has long been attributed to the apocalyptic "Four Horsemen," Death, Famine, War, and Pestilence. These can be translated into ecological categories that are equally distinctive: death as the aging and intrinsic deterioration of the individual; famine as the limitations in available energy; war as competition within a species for space, energy, and mates; and pestilence as competition between species, particularly by parasites and predators, for the resources of the community.

Age Levels

Regardless of their initial mortality rate, the members of most species in each natural ecosystem exhibit a stability of age structure that has yet to be explained or even duplicated in a laboratory. The rates of natality and mortality match closely, and environmental factors balance out whatever fluctuations arise from year to year. Some species are represented by only one age group at a time, as is usual among annual plants and most kinds of insects. Others have annual breeding periods but live for several years, producing a population composed of several year-classes or cohorts whose lifespans overlap. The year-classes can be counted separately, as gamekeepers do in monitoring the welfare of deer or pheasants.

Plants and animals that live for many years and reproduce repeatedly may form populations like those of Norway rats and human beings, which breed continuously, obliterating the signs of annual additions. The welfare of these populations can best be recognized by examining the trend in proportions of prereproductive, reproductive, and post-reproductive numbers as shown in an "age pyramid."

A pyramid that is excessively wide at the base and narrow at the top shows a population in peril, increasing rapidly because natality far exceeds mortality; lack of food or space or both are likely to be critical. It is the type of population that Thomas R. Malthus, an English clergyman and economist, feared when he wrote his 290-page book *An Essay on the Principle of Populations*. This volume, which influenced the thinking of both Charles Darwin and Alfred Russel Wallace, was first published anonymously in 1798 because the dire predictions of famine and conflict made the author fear the response to it.

In the United Kingdom today, the age pyramid is quite different, for it is undercut at the bottom due to diminished natality. Progressively the reduction rises to older levels, slowing the rate of population

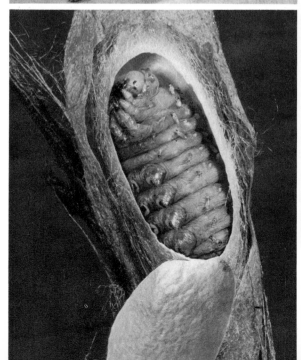

With thousands of different insects going
through regular transformations from egg
to active immature, perhaps to inactive
pupa, and then to dispersing adult, the
environment is full of examples of
different types of reproduction. A moth
lays her eggs in an irregular row beneath
an aspen leaf (above). The male water bug
Zaitha waits until his mate has cemented a
whole raft of upright eggs upon his back,
and then moves about in the pond as a
living perambulator (right, center). The
larvae of the little fruitfly Drosophila,
so much studied in genetics, attach
themselves for pupal transformation on
some firm surface (top, right). The
cecropia moth caterpillar spins a tough
cocoon of silk from glands near its
mouth, then gets ready within its shelter
to shed its skin and expose the pupal
covering (right). The monarch butterfly,
like most butterflies, suspends itself as a
naked pupa (called a chrysalis) until it is
ready to emerge and hang inverted while
its wings expand and dry (far right).

growth when the trend reaches the reproductive ages. This time lag is the reason why any change in human reproductive rate takes 20 to 30 years before it affects the population trend. The effect comes sooner in ethnic groups and nations in which the interval between one generation to the next is small. It is slower in the United States, since 60 per cent of the children are born to women between ages 20 and 29, not 14 to 25 as in India. The recent record low in the birth rate in the United States is due in large part to the smaller proportion of women in the child-bearing age group as compared to the Depression and World War II years (1930–1944), when families were unusually small. The rate will rise inevitably as soon as the "war babies" who were born after 1945 reach their peak of productivity. Even if they bear no more than one child apiece, and not the two that are ordinarily thought of as basic to Zero Population Growth, their numbers are great enough to continue the population rise.

Over most of the world, the human birth rate exceeds the mortality rate by at least a factor of two, and the age pyramid has a wide base. For 1969 the Population Reference Bureau indicates that the rates per thousand were:

Already people are essentially committed to the program of population rise that will continue to 2000 A.D., barring a real catastrophe of some kind. The forecasts which the United Nations publishes every year now are concerned with changes in the proportions of the total that can be expected continent by continent. African and Australian populations are increasing at close to the average rate for the world, and can be expected to maintain their relative proportions in the total. A major increase is seen in Asia (excluding the U.S.S.R.), and a smaller increase in Latin America. Corresponding minor decreases are expected in Anglo-America, and a major shrinkage in Europe and the U.S.S.R.

Density

The prediction for the year 2000 is 7,522 million people if no significant progress is made in lowering the birth rate. Of these people, 62 per cent are likely to be Asians, crowded into 21 per cent of the world's habitable land—nearly eight times as many as lived there in Malthus' day. Spectacular feats of food production and distribution seem necessary immediately to stave off strife and famine. So far, hopes for the "Green Revolution" have proved impractical. Competent observers predict calamity

Population increases per thousand in 1969

Country	Births	Deaths	Population increase	
Ceylon	32	8	24	(2.4%)
Mexico	43	9	34	(3.4%)
Sweden	15	8	7	(0.7%)
United States	17	9	8	(0.8%)

starting in Asia long before the end of the century. They disagree mostly as to whether it will come in 1985 or 1975.

The density of any species is the number of individuals on a unit of area or in a unit volume of habitable space. It gives a measure of the competition for energy and territory. In the sea it relates to the available share of dissolved nutrients, and on land to the fresh water that can be reached. It tells much about the distance that a disease organism or a predator must go to the next potential victim. All of these factors influence the rate of mortality. Many of them affect the young more than individuals whose growth has been completed. For this reason they mold the age pyramid and affect the survival of the species.

A measure of the density of a species is like other statistics in having a limited value until a great deal is learned about its applicability. The details of the life history are particularly significant. A thousand frog tadpoles per cubic meter of water in a shallow pond on April 15 gains relevance when followed by information on the number a week later. But neither number shows whether the organisms are uniformly distributed or clumped—perhaps close to a food supply. Clumping is more usual among living things than uniformity, if only as groups of offspring clustered around a parent, or as offspring in one region (such as young lobsters among the plankton) and adults in another.

To avoid being distracted by numbers

Density of major herbivorous mammals in Kruger National Park

Species	Total number	Mean weight in pounds	Density per square mile	Biomass per square mile	Percentage of herbivores
Impala	204,050	90	27.72	2,495.1	23.70
Elephant	2,374	7000	0.32	2,257.9	21.45
Cape buffalo	10,614	1100	1.44	1,586.3	15.06
Hippopotamus	2,865	2500	0.39	973.2	9.24
Zebra	14,400	475	1.96	929.3	8.83
Blue wildebeest	13,035	400	1.77	708.4	6.72
Giraffe	2,975	1500	0.40	606.3	5.76
Kudu	6,875	380	0.93	354.9	3.37
Waterbuck	4,085	450	0.55	249.7	2.37
Others	23,079	150	2.44	368.5	3.50
Total Census	284,342	273	38.59	10,529.6	100.00

Others include : white rhino, eland, roan antelope, sable antelope, tsessebe, nyala, reebuck, steenbuck, Sharpe's steenbuck, duiker, red duiker, warthog, bush pig, bushbuck, mountain reedbuck, klipspringer, suni, and oribi.

alone, ecologists have devised a concept that encompasses both the number of individuals and their average size, species by species within an ecosystem. A million bacteria could form a thin film over a small area. A million elephants may be more than the world has ever supported in a single year. By multiplying the number of individuals in a region by the average weight of living cells per individual, the ecologist arrives at a measure he calls the biomass. Generally it is expressed in pounds per acre (or square mile = 640 acres) or in kilograms per hectare, where 1.12 pounds per acre equals one kilogram per hectare.

South Africans recently made an aerial survey of the large herbivorous mammals in their Kruger National Park. The nine commonest species were found to account for all except 3.5 per cent of the biomass at this trophic level on the 7,360 square miles within the boundaries. (See chart on page 121.)

For convenience, the weight of an animal is generally used in computing the biomass of the population. Properly this weight should be adjusted to avoid including any nonliving skeletal materials, the content of the gut and bladder, or any other parts that are not live cells carrying on respiration and using energy. The biomass of plant material should not include cellulose cell walls, resins, and other nonliving materials for the same reason. Corrected in these ways, the biomass becomes a particularly meaningful way to relate the species in each ecosystem into a pyramid that reflects accurately their energy relations. As would be expected, the biomass of producers usually exceeds that of herbivores, and that of herbivores the biomass of carnivores. More surprising is that, at each trophic level, large animals tend to maintain a larger biomass than any

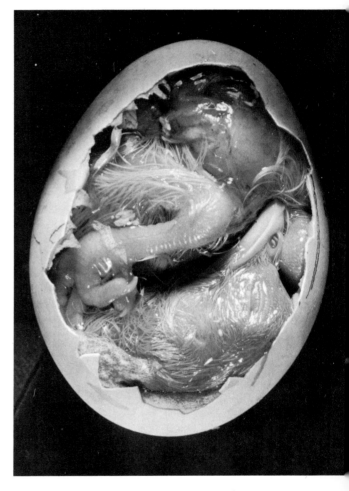

single kind of small animal. Diversity evens out this feature, for each ecosystem has only a few kinds of large animals and many species of small ones.

Sometimes the important feature in a trophic level is the size of the "standing crop" from which members of the next-higher level get their energy. Standing crop is not equivalent to biomass because it excludes the many individuals that must survive to reproduce—at least until they have done so. On the other hand, it includes nonliving organic matter such as hair, lipid

Man must now make value judgments in coping with expanding populations. For thousands of years swarms of grasshoppers (right) in North Africa and the Near East have devastated crops. Now an International Locust Commission is seeking to control them, although doing so requires more cooperation than many small countries seem able to give. In man's major cities, equivalent chaos is temporarily avoided by housing dense human populations in enormous apartment houses (overleaf, top). In contrast, on San Martin Island, off the Peruvian coast, the human population may be just one man (see center of photo), while

droplets, chitin, and cellulose, because these can serve as nutrients to some consumers and decomposers. Snail shells and limy skeletons, however, yield no energy and are properly excluded from measures of both the standing crop and the biomass.

Regulation of Population Size

From observations in the field and tests in the laboratory, ecologists have come to expect the population of every living thing to follow a characteristic pattern of growth. Initially, as when a pioneer reaches a new site with no obvious competitors, or a microbe is introduced into a tube of culture medium, the resources appear limitless. The organism is free to express its full biotic potential. The population of descendants increases at an ever-increasing rate, in the "geometrical" progression that Malthus recognized, such as one microbe becoming two, the two four, the four eight, the eight sixteen, and so on. A rise of this kind is now generally referred to as logarithmic or exponential.

Before long, the increasing population runs into environmental resistance. We can think of the organisms as having saturated their habitat. The ecologist speaks of the population as reaching the carrying capacity of the environment, and looks for the factors that tend to regulate the numbers of individuals. Ideally these factors prevent the population from destroying its resources.

The biotic factors that regulate the size of a population are clearly related to density, hence density-dependent. They include competitors for food, parasites, and predators, and even the availability of mates. As any population grows, each individual tends to get a smaller share of the limited supply of energy. With malnutrition, the organism becomes more susceptible to diseases and harmed by parasites; at the same time, high density of the host organism facilitates the spread of parasites. Predators may be less significant than is generally supposed, mostly because they tend to multiply more slowly than their prey, to live longer, and to be subject to starvation in off-seasons as the physical environment limits the productivity of the ecosystem. Potential mates are most numerous at high densities, but so are interruptions that may prevent successful mating; competition for space may leave few areas of privacy in which eggs or young can be given sanctuary. Repeated encounters with other individuals in the population may also sustain a high level of stress, which acts to prevent reproduction. Stress is not infectious, yet it often acts like a disease in undercutting the age pyramid and causing a dramatic decrease in the numbers of a species.

From his extensive studies on animal populations, V. C. Wynne-Edwards of the University of Aberdeen concludes that in most natural communities, starvation is rare, and neither predators nor disease are primarily responsible for keeping the density of a species reasonably constant. He suggests, instead, that each kind of animal has evolved some type of automatic restrictive mechanism, which operates with an efficiency equal to that of territorial behavior in birds or hierarchical systems in birds and mammals. These inherited patterns of social action effectively keep a large fraction of the adult population unemployed in reproduction, and sustain a balance between natality and mortality that corresponds closely to the carrying capacity of the environment. Some other animals, including foxes and certain kinds of rabbits

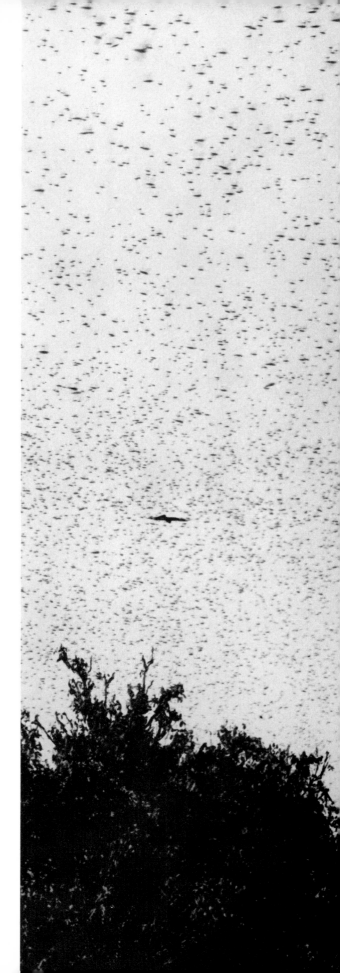

the cormorants known as "guanay birds" compete for nest sites. The scarcity of sites is a limiting factor to the cormorant population (overleaf, bottom). Man faces similar problems: on the island of Hong Kong, every hillside is dense with the shacks of the poor and the streets teem with people (second overleaf).

and deer, react to crowding either by a decrease in the rate of ovulation (due to a lowered secretion of sex hormones) or resorption of embryos after they reach the uterus.

None of the density-dependent interactions may save a population from disaster if it is isolated and all major predators are removed consistently over a long period. Ecologists learned this from a program that has now become a classic in wildlife management. It began in 1906, when President Theodore Roosevelt created the Grand Canyon National Game Preserve to protect a particularly fine herd of mule deer in northern Arizona. He was correct in believing that these animals were facing stiff competition from about 20,000 cattle, 30,000 sheep, and a large number of horses on the magnificent Kaibab Plateau, 90 square miles that are isolated from the rest of the country by Grand Canyon on the south, deep but lesser canyons on the east and west, and the semi-arid lands of Utah to the north. But when the stockmen and their domestic animals were removed in 1906, the program for elimination of all major predators was intensified "to protect the deer herd." Already the stockmen had been warring against predators for nearly 20 years. Between 1906 and 1931 another 816 cougars, 30 wolves, 7,388 coyotes, and 863 bobcats were eliminated on the plateau by shooting and trapping.

By 1918, the Forest Service reported damage to young trees by deer on the Kaibab preserve; the herd was estimated to include 40,000 animals—one deer for each 17 acres. In 1923, the famous naturalist George Shiras III visited the plateau, and implored the Forest Service to cull the herd, claiming that 30,000 to 40,000 of the animals were on the verge of starvation.

By autumn of 1924, a census showed about 100,000 deer—one per 7 acres. Licensed hunters removed 1,000 of them; an estimated 60,000 died of starvation the following winter. Again permits were issued, and in 1925 another 1,000 deer were shot. Before spring of 1926, 60 percent of the herd had died for lack of food. Two years of winter kill accounted for some 96,000 deer, where hunters under permit took only 5,000 in five consecutive years. Despite a legal battle between the State of Arizona and the Forest Service over the right to issue hunting permits, a program unpopularly known as "government killing" was instituted, decreasing the number of surviving deer by 1,124 in 1928, by 4,400 in 1929, and by 5,033 in 1930. This brought the herd to an estimated 20,000 animals. The number was still excessive because the starving animals had done so much damage to the vegetation on the isolated plateau, shrinking the carrying capacity of their environment. Since 1939, the herd has been held at about 10,000 animals—one for each 68 acres—and the Kaibab has recovered much of its original charm. (See diagram on page 132.)

Physical factors in the environment tend to control the size of a population without reference to how many individuals are present. They are density-limiting, but density-independent. This category of hazards includes prevailing weather which determines in temperate and polar lands the length of the growing season and the ultimate availability of solar energy. Space or shelter is a physical factor too, yet one that operates through the biotic factor of competition. Hence it is density-dependent, becoming critical as the size of the population increases.

Ordinarily, immigration from more populous areas tends to hasten the day when a population will reach or exceed the carrying capacity of its habitat. Similarly, emigration provides a safety valve for the population that has already fully exploited the standing crops of its resources and is in danger of rising further in density and destruction of its environment.

Quite often, wildlife managers overlook immigration until a population grows excessively. This happened recently in Africa's Kruger National Park, where the history of the elephant herd was well known. When the sanctuary area was set aside in 1905, it held fewer than 10 elephants of all ages. They were the elusive individuals that somehow had escaped the hunters who wanted their tusks. Despite poaching in the park, the greater security let the herd increase to about 500 in the next forty years. By 1960, the population reached 1,000. The park managers were delighted. Their report includes the estimate that immigration of elephants into the park from the adjoining wild country across the border in Mozambique and Southern Rhodesia probably equalled the annual loss of 30 to 50 elephants through continued poaching and the occasional emigration of an elephant out of the park into South African farmlands. The managers urged that a stronger fence be built to halt immigration, and that between 100 and 150 elephants be removed annually to compensate for the normal increase due to births.

A study of the feeding habits of adult elephants revealed that each one eats daily about a ton of selected shrubs, tree foliage, and herbaceous plants. To support an elephant, about three square miles of park are needed. If the animals dispersed themselves uniformly, a total of 2,440 of them

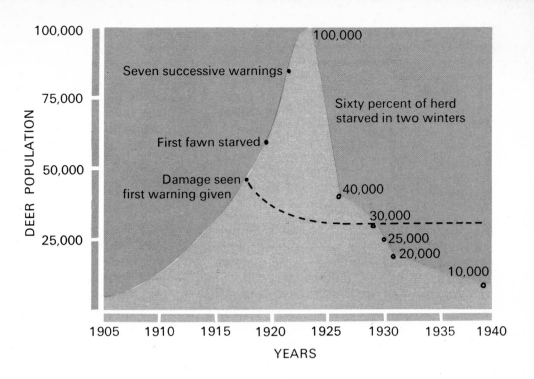

might be accommodated. Any more would exceed the carrying capacity of the environment. Yet, in the face of increased immigration, the deliberate removal of elephants of both sexes and selected ages failed to keep the population at ideal levels. An aerial census in 1964 showed 2,374 elephants in the park. A repetition in 1968 revealed 6,093 of them. By October 1970, the number rose above 8,000. Far more elephants were breaking the fence and entering as immigrants than had been imagined.

For any population, conditions tend to be most stable near the center of the geographic range, where physical factors are most favorable and the organisms live in their preferred habitat. The population is least stable around the extreme periphery of its range, where one factor or another causes physiological stress. The periphery is the departure area for emigrants and the arrival zone for immigrants.

Both emigration and immigration are density-dependent phenomena. Ordinarily they are studied in populations that are undergoing major fluctuations. Best known are those of grasshoppers ("migratory locusts") in arid lands and of lemmings on the arctic tundras, where they are the commonest rodents. Both of these types of life undergo behavior changes as their populations grow. The grasshoppers (*Lo-*

custa migratoria and others in the Old World, and *Melanoplus mexicanus* in the New) remain sluggish, nongregarious and nonmigratory so long as each immature individual during its early stages of development sees few others of its kind. But when the density of the population rises in a region, letting young insects of these species see neighbors in fair numbers, metabolic rate and behavior are affected, leading to synchronous flights of hungry adults.

These extreme fluctuations in animals that can move about freely are typical of ecosystems that are relatively simple because of chronic physiological stress. The arid area where locusts develop and the tundra where lemmings live both suffer from drought, in the one case from the excessive drying power of the warm air, and in the other from prolonged unavailability of water due to cold. Few species have the tolerances required to survive under these conditions, and the food web has a minimum of alternative pathways.

Buffer species are essential to give an ecosystem stability and reduce fluctuations in its component populations. When one kind of food plant is depleted or a herbivorous prey becomes scarce, the animals that use the particular form of nourishment need another to which they can transfer

The first major test of predator control and an isolated population of large herbivores occurred on the Kaibab Plateau in northern Arizona. The mule deer herd increased far beyond the carrying capacity of its food supply when 600 mountain lions were removed in 1907–1917, 74 in 1918–1923, and 142 in 1924–1939.

Concurrently, 3,000 coyotes were removed in 1907–1923 and 4,388 in 1923–1939. After 11 wolves were destroyed in 1907–1923, the few that remained were eliminated by 1926. The dotted line shows the probable capacity had the herd been reduced in 1918.

their attention while the first recovers. The more complex an ecosystem is in terms of species structure, the more it resists change and the more it challenges scientists to understand it. Conversely, the fewer species an area supports, the more vulnerable it is to violent fluctuations.

The ultimate in simplification is man's invention: monoculture. In unfenced, unroofed areas measured in the millions of acres he strives to raise a single crop of food or fiber for his sole benefit. Species from the surroundings invade and multiply rapidly, becoming weeds or pests according to whether they compete with his crop or get energy from it. Man attempts to defend his investment through programs of clean cultivation and chemical treatment. They eliminate the buffer species, frequently forcing native organisms to become pests for a lack of preferred foods. Nor is a reservoir of predators, parasites, and agents of disease left to slow the rise in numbers of a pest until the physical environment can bring relief.

Recently the Canadian entomologist Philip S. Corbett equated a complex ecosystem, rich in buffer species, with environment of high quality. He questioned how far man could afford to simplify the ecosystem, reducing its quality, before the cost in outbreaks of pests and chemical pollution exceeded any gain in yields of crops. Dr. Corbett identified each outbreak as an indicator of oversimplification. He pointed out that a small number of the pests could be tolerated; irruptions were intolerable. To seek an environment completely free of pests would be uneconomic. To make a cropland outbreak-free would require informed manipulation toward restoring a measure of biotic diversity. He believed that the yield from the crop would not be significantly reduced over the long term by a consistent program aimed at preventing outbreaks.

In studying irruptions, scientists have learned how nearly impossible it is to deflect or destroy the pest population once the outbreak begins. But when it has run its course and subsided, as each outbreak does, efforts can profitably be turned toward preventing a repetition. Regulation of the populations of potential pests by biological (rather than chemical) methods offers real hope. Resistant crops can be chosen, even if their yield is slightly less. Newer and more sophisticated techniques have been tested. Synthetic attractants, made almost identical to those that many female insects use to guide males to them, may be set out, either to lure males into traps or to so confuse them at mating season that a majority of females go unmated. New applications are being tried of an effective method that was used recently in the southern United States when an outbreak of an introduced screwworm fly threatened the cattle industry; it consists of raising and freeing at the right time and place of large numbers of males that have been sterilized with radiation; they mate with normal females, whose eggs then fail to develop; in a few generations, the pest population shrinks to zero.

The time to act is often before a pest recovers from adverse weather, such as winter or a prolonged drought. Over much of Australia, where the introduced South American disease known as myxomatosis is no longer so effective in curtailing the destructive populations of introduced European rabbits, wildlife officers are now benefitting from dry periods. A few years with little rain reduce the rabbits to isolated colonies, which often can be spotted from the air and exterminated by men on the ground, leaving no survivors as a breeding stock toward another outbreak when the rains return.

Mankind Multiplies

Experience with the growth of nonhuman populations has given ecologists fresh ideas about the increasing numbers and rising expectations of people. Studies made in the years since 1950 have convinced many scientists that the carrying capacity of the world for mankind matches a world population no larger than 500 million. This total was reached soon after colonists from Europe began settling in the New World. Since that time the human population has grown about sevenfold, and is likely to double again in less than 30 years. Now the deleterious effects are increasingly obvious in the environment. Mankind has become a species in an outbreak or epidemic phase, and a major ecological problem.

Human ecology has emerged as the fundamental study in the United Nations, which was incorporated in 1945 primarily as an institution to foster world peace. Quickly it became a clearing house for international information about human populations, food resources, health, and technical development. Raw data for its Statistical Office were gathered by agencies known as FAO (Food and Agricultural Organization), WHO (World Health Organization), and UNESCO (United Nations Educational and Scientific Organization).

For the year 1950, the United Nations obtained the most reliable estimate of the world's human population achieved to that date. It also examined pertinent information from the past, and provided a new perspective on the proliferation of mankind.

Not only was approval given for an estimate of 5 or fewer million people in 10,000 B.C., but totals were suggested for 1650 A.D. and every half-century thereafter. Calculations from the data gave approximate proportions of the world population on each continent, and the trends of change in recent times.

The carrying capacity of the land increased soon after 10,000 A.D. for a portion of the total population, wherever people transferred from hunting and gathering food to the deliberate cultivation of domesticated animals and plants. Slowly the population increased, to about 545 million by 1650 A.D. This represents between 6 and 7 doublings of the total, once for each 1,712 years on the average: 5 to 10, to 20, to 40, to 80, to 160, to 320, and about eight-tenths of the way to 640 million. Probably this required no change in the birth rate, which is around 50 per 1,000 of population in primitive societies. The mortality rate must have been only slightly below this figure, and the average human lifetime between 25 and 30 years.

A scattering of records that seem reliable confirm that these were the facts of human life in Shakespeare's day. In the parish of St. Botolph, which then was near the edge of London, England, for each thousand babies born alive, 61.5 were born dead, and 139.6 died within their first month. Of the surviving babies, who were christened and given a name, about 220 died before age 5. Fewer than 300 out of each thousand lived to age 15. Yet a few people survived all of the hazards that the sixteenth century offered. Several lived to become centenarians.

During the 1600's, the mortality rate diminished noticeably. More people lived longer, and the population increased at a faster rate. Many factors contributed to this change, among them the large-scale emigration from Europe which prevented any major increase in density as the total numbers grew. At intervals, wars and famines and epidemics of disease killed thousands of people. Despite these setbacks, the population of the world reached 728 million by 1750, 906 million by 1800, and 1,171 million by 1850. The new rate of doubling was once in only 186 years—almost ten times as fast as in the preceding periods.

Although so many emigrated to the colonies, the number of Europeans who stayed home increased from 18.3 per cent of all humanity to 22.7 per cent. Working and reproducing, they became processors of raw materials from other continents and importers of food for their own needs. Yet the penalties of monoculture appeared too. An alga-like fungus (*Phytophthora infestans*) spread to the European continent and then to Ireland, destroying the potato crop upon which so many small families had come to depend. During 1845 and 1846, disastrous famines in Ireland led emigration to the United States alone to jump from about 4,000 annually to 61,242 in 1845, 105,953 in 1846, double this in 1847, and a steady exodus at a lower rate for the rest of the century.

During the two centuries that ended in 1850, immigration (mostly from Europe) accounted for much of the 430-per cent gain in the human population of Anglo-America, from a mere 0.2 per cent of the world total (virtually all Amerindians) to 2.2 per cent. Latin America showed a gain from 2.2 to 2.8 per cent, while Asia changed from supporting 60.5 to 63.9 per cent, without much mobility in the population. In this race toward increased human numbers, Africa and Australia fell behind.

For the year 1950, the world population lay between 2,350 and 2,471 millions, with 2,400 millions regarded as a "good round number." The total had more than doubled in a century. Birth rates in Western countries declined—in Scandinavia, for example, from about 32 per thousand in 1850 to 15. The mortality rate fell faster, due to modern medicine and better programs of public health. In other parts of the world, the economic and sociological changes that led to the lowering of birth rates in the industrialized nations have never come. Birth rates remain high, and the population increases rapidly because mortality has been reduced through import of medicines, the new insecticides, and large amounts of donated food. The undeveloped countries suddenly have more young people surviving, to need food, education, and employment for which local resources offer little.

The United Nations' reports on world population growth since 1950 leave no room for doubt that the doubling time for mankind has shortened to about 37 years. By mid-1962 the total reached 3,115 millions, and 3,750 by mid-1970. The population curve continued to rise toward about 37,000 million a century from now. Almost no one really expects this number to be reached. But when the curve will reverse its form and end the outbreak remains as unpredictable as the force that will cause the change.

9 The human environment

The environment that each person encounters during his lifetime changes far more than one generally realizes. We easily forget that the first 40 weeks, before birth, are spent in an aquatic situation. The mother's genetic makeup and outer environment influence the embryo's development. What she breathes, eats, and drinks, and what infections she contracts, affects the child's life more before birth than subsequently.

The habitat of the unborn child has certainly changed far less during the past two million years than the world that each baby enters at birth. The normal environment inside the mother is ordinarily well matched by the pattern of growth that each human embryo inherits. Yet only recently has scientific inquiry turned toward understanding the interaction between parent and young during those first 40 weeks.

Today it seems strange that for more than two centuries the scientists of the world possessed and used the microscopes that are needed for viewing fertilization before anyone actually discovered the event. The first person to see a sperm cell penetrate an egg was Hermann Fol of Geneva, Switzerland, in 1879. Until less than 70 years ago, no one found or was able to isolate a live human egg, although each is about 6/1000 of an inch in diameter, and quite visible to the normal unaided eye.

Fewer than 500 eggs mature in a woman's lifetime, out of many thousands of potential candidates in her ovaries. Each egg ruptures from its follicle on a schedule that recycles approximately every 28 days. The egg is carried by a gentle stream of watery fluid produced in the body cavity. This fluid flows through the Fallopian tubes to the womb, where it generally is reabsorbed completely into the blood.

The tiny rivulet of clear liquid from the woman's body cavity is the normal environment in which a sperm cell meets an egg, and for the earliest development of a new human life. Experiments conducted at the Wistar Institute of Anatomy in Philadelphia and elsewhere show how unpredictable that meeting is. As each sperm swims on its own, it depends upon the waterway for guidance, support, and nourishment. It has no appreciable store of energy with which to propel itself so many thousands of times its own length. Its payload, which fills its head, is a big nucleus containing all of the hereditary material. At the tip is a minute bag of enzymes with which the sperm can make an entrance into the egg. In the base of the lashing tail is the enzymatic machinery needed to release useful energy from the organic nutrients that the sperm absorbs from its watery environment.

We can guess that the sperm uses mostly glucose—blood sugar—as its propulsive fuel. But the amount of glucose available from its surroundings is never great. The sperm may absorb also at least half a dozen different amino acids in trivial amounts, and convert them enzymatically for service in the same energy-releasing cycle. To measure these events in so minute a cell is far more difficult than to demonstrate a change in the swimming rate and in sperm survival when the acid-alkaline balance of the fluid environment varies from the normal range. Probably the swimming rate reflects the pace at which nutrient materials can be absorbed and mobilized. The longevity of sperm cells is as full of mysteries as the aging process in the body as a whole.

All further development is contingent upon the admission of the nucleus from a sperm into the egg. Only then will the

At a cost of nearly 100 human lives and many millions of dollars, the huge Hoover Dam of poured concrete was built across the Colorado River. This inundated 247 square miles of valley land and drowned all of the native, *well-adapted desert life that could not move elsewhere to uncontested areas. The dam created Lake Mead as a reservoir of water for production of hydroelectric power and as a recreation center.*

single cell possess a double set of genetic instructions, and begin to follow them. The chromosomes that contain the inheritance perform a special ballet that intermingles 23 from the sperm with the 23 from the egg. The action slows slightly while each threadlike chromosome separates into a lengthwise pair of strands. Then a representative of every chromosome from both parents—46 in all—is whisked into each of two clusters. Between them the boundary membrane of the fertilized egg folds in to make one cell into two. Succeeding divisions are simpler and faster. By the time the new embryo wafts out of the Fallopian tube into the narrow confines of the womb, it is already a cluster of cells in a pattern that is recognizably mammalian.

137

Two days may suffice for the travels of an egg through a Fallopian tube. Thereafter the embryo is in contact with a moist surface, the cells of which have already responded to some of the hormones in the mother's blood. These wall cells of the womb have dissolved foods to offer and are so responsive to contact with firm objects the size of a fertilized egg that they will enfold an inert glass bead of these dimensions.

Interaction between the newcomer in the uterine confines and the uterine environment works both ways. Chemical signals, that have not yet been identified, diffuse from the developing embryo into the uterine wall, stimulating it to proliferate and extend its network of fine blood vessels. An unfertilized egg produces none of these signals, and the wall cells then secrete digestive enzymes instead, which liquefy the dying egg and absorbing its chemical residues.

The critical point in the development of a human embryo comes when it is nearly three weeks beyond conception and less than an eighth of an inch in diameter. For its rapid growth it needs more food than comes to it by diffusion from the dilute liquid in its recess in the uterine wall. Already the embryo has a heart and some blood vessels. It organizes a circulation through an external fold of its body, one that is specialized to pick up nourishment from the mother and transfer dissolved wastes to her blood stream for her to discharge. The specialized fold, known as the chorio-allantoic membrane, digests its way through the uterine wall, right into maternal blood vessels. How the mother tolerates this intrusion of fetal tissues without developing fatal blood clots has yet to be discovered. For about eight months each of us makes our mother's flowing blood our warm, nourishing environment.

Infection can alter the chemical nature of a woman's blood stream and affect the development of a fetus. If she contracts rubella (German measles) during the 4th to 9th week of pregnancy, the virus may damage the heart of the embryo; during the 5th to 8th week, the lenses of the eyes are vulnerable; and from the 7th to the 12th week the inner ear mechanism, upon which normal hearing must eventually depend, may be injured beyond repair. A number of bacterial infections, most notably syphilis and typhoid fever, can cause serious damage to embryonic tissues if contracted by the mother during the first 3 or 4 months of pregnancy. Nor is the unborn child the only endangered partner in this relationship before birth: the virus of influenza is a far greater hazard to the life of a pregnant woman (and to the fetus, if she dies) than to nonpregnant people of the same age.

What a change we all face when our mother's tolerance for us ends and, with labored contractions of her uterine muscles, she shoves us out into the thin, non-buoyant air of an unfamiliar world. So far, no one has discovered how much advance warning the fetus gets. The rhythmic contractions may force it from its private aquarium—the amniotic fluid in the "bag of waters"—into a dry world of variable temperature at any time between the 28th and 44th week of normal development.

The Newcomer

Once born, the baby must breathe, digest, and excrete for itself, as well as nurse and yell. Quickly its blood stream follows a new course, bypassing the old umbilical arteries and veins that no longer connect to a placenta embedded in the mother. Its nervous system takes note of a broad spectrum of sights, tastes, smells, contacts, and temperature changes as the learning experience begins. Sounds come more clearly than they did through the mother's body. For every one of these changes the baby must have been prepared by its previous growth. No future shift in its environment will be so swift and drastic. Exposed to dry air, the skin of the newborn infant responds noticeably within a few hours. Its outer layers, which are composed of cells already dead but not yet sloughed away, lose moisture and toughen. They are the first line of defense between the live body and the surrounding world.

Initially, the wastes voided from the baby's digestive tract have no relation to its new food. Instead, they are residues from its previous life within the womb. Stained green with bile pigment and suspended in mucus, they are mostly the remains of dead cells from the skin. The pigments are green because they have had no encounter with bacteria that could oxidize them to a brown color. The dead cells are those that were lost into the amniotic fluid surrounding the unborn infant, and then swallowed as the baby drank from its liquid environment. Ordinarily this automatic swallowing keeps the hydrostatic pressure around the baby in the proper range by removing liquid and transferring it by way of umbilical arteries and placenta to the mother's circulation.

Born without bacteria, the baby begins immediately to acquire these potentially dangerous companions over its skin, on the linings of its breathing passages, and from end to end of its digestive tract. At first it must rely upon the protective action of the antibodies its mother produced in response to these same kinds of bacteria, and transferred through the placenta into the blood stream of her unborn infant. Yet almost at once, the baby begins making its own antibodies, building active immunity to the potential invaders that surround it on all sides.

In its bone marrow, the infant responds noticeably within a week or two by producing the adult type of hemoglobin in new red blood cells. Although the color of this pigment is essentially the same as that of fetal hemoglobin, the molecular details differ. They increase the efficiency with which oxygen can be picked up in the fine blood vessels of the lungs and released in blood vessels close to the body tissues. This response to getting oxygen through the lungs, instead of by way of the placenta, takes progressively greater advantage of the easier access to the vital gas from the inhaled air.

Week after week as the baby nurses, the chemical qualities of the mother's milk change. They match the development of digestive and absorptive capacities in the infant's alimentary tract. Yet the formula called for by the inheritance of the mother is modified consistently by the tissues of the mammary gland according to a single pattern that has proved satisfactory with thousands of generations of children in the past. A mother kangaroo, by contrast, can sustain two patterns at the same time—one in the glands supplying nipples to which she has young attached within her pouch, and a different constitution (richer in fat) for the offspring that have left the pouch but still return for nursing on free nipples.

Seldom does a child make the choice to change from a diet of milk to one of other foods. Weaning is usually a decision of the mother, and she expects the young digestive tract to cope with unfamiliar materials,

Irrigation water helps man raise cotton on a broad, almost level valley in Eloy, Arizona (top, left). A massive dike in the Netherlands protects manmade agricultural land from the sea that once covered this whole area (bottom, left). With a crude plow pulled by a water buffalo, a Javanese farmer (below) prepares his flooded rice paddy for the planting of another crop.

which differ greatly according to geography and culture. In many parts of the world, this is the time in development when the environment ceases to supply enough protein. Chronic protein deficiency leads to many symptoms of retarded growth, both physically and mentally. The pattern is known by the name kwashiorkor, a West African word for a weanling child. In many warm parts of Asia, Africa, and Latin America the symptoms are often thought of as normal because they are so nearly universal. A deficient environment of this kind affects chiefly the poor and ignorant in Europe and Anglo-America, in families where quality of food is less a goal than a sufficient quantity.

The Social Environment and Learning

As the human child grows, the relative importance of information from its environment changes. Experience adds to the memory bank of visual configurations that can be identified, increasing reliance upon sight. Taste centers in the throat, cheek linings, and palate degenerate, leaving only those on the tongue to distinguish combinations of sweet, sour, bitter, and salty. The ear loses its sensitivity to high-pitched sounds at 20,000 cycles and above. Improvements in reflexes called for through our sense of balance come through further development of the nervous pathways as well as practice. Development follows an inherited schedule, which determines more than anything else when a child can learn to walk and when it will pass the toddling age.

Delicacy of touch diminishes as the skin thickens, but never loses importance. Stimulation from physical contacts with parents and at least some other individual of the same general age seems indispensable to the normal development of social behavior. A child that is fondled can adjust to a congenital handicap, such as blindness or deafness, whereas a neglected child with good eyes and hearing has social difficulties approximately in proportion to its past deprivation of physical contacts.

Learning holds new importance as we reach full size and our skeletons lose some of their juvenile flexibility. By then our nervous control, with fast reactions to each stimulus, and our muscular strength may attain their peaks of performance. With luck we attain a fair measure of our personal capacities by the time we free ourselves of parental guidance and feel the full impact of diverse challenges from the physical environment. We need this measure of ourselves as a basis on which to estimate the risk in each opportunity we recognize, and to judge what course to follow. Memories of successes and failures become important. They tend to steer us from accidents that could happen easily, and do in moments of carelessness. In consequence, our performance often improves despite signs that age is slowing our responses and limiting our physical prowess. So far as is known, no other kind of animal benefits so much from experience or extends its life significantly in this way.

Learning lets us cope with an increasingly unnatural environment of man's own making. We benefit from warning signs and traffic signals, and include in urban planning a great many safeguards against dangers that are foreseeable. Today we see indications of peril from accumulating wastes, and interpret them as requiring a change cutting across cultural patterns that have been evolving for millennia. The change would wean us forcibly from a familiar way of life and death, whereas a continuation without change seems sure to destroy the ecosphere.

The First Environments of Man

It is only a century since Charles Darwin's expected but highly controversial book *The Descent of Man* was published (1871). Darwin's friend, the outstanding zoologist Thomas H. Huxley, felt obliged to state before an assemblage of zoologists that

> Whatever system of organs be studied, the comparison of their modifications in the Ape Series leads to one and the same result—that the structural differences which separate Man from the Gorilla and Chimpanzee are not so great as those which separate the Gorilla from the lower Apes.

During the intervening years, scientists have found a vast body of supplementary information which supports the same conclusions, based originally on anatomical resemblances, vestigial organs, and embryonic homologies. With far more confidence and accuracy, they can consider the environments in which the human species arose and how both these and mankind have changed.

In Darwin's day, Neanderthal man was the only fossil primate known—and not well known, at that. Modern paleoanthropologists suspect today that, although Neanderthal man was on the line of descent to the mongoloid, negroid, and caucasoid peoples of the present, there were never more than about 1,500 generations of Neanderthalers, give or take about 250 generations. Modern techniques applied to the fossil remains and associated artefacts allow assignment of dates to Neanderthalers: from slightly prior to the last glacial period, about 75,000 ($\pm$5,000) years ago to around 35,000 ($\pm$3,500) years before the present.

The oldest human remains so far discovered are from the Olduvai Gorge in Tanzania, excavated by Dr. L. S. B. Leakey and his scientist wife, Mary Leakey. But whether to call these fragments *Homo*, or some other genus, remains controversial. No longer does the decision turn on whether pre-men became men when they learned to make and use tools. Apparently several different kinds of manlike or human primates evolved this ability. Chimpanzees carefully cut plant materials as tools. The cactus finch in the Galápagos is a toolmaker and user. The Egyptian vultures use stones as tools for opening ostrich eggs, and some solitary wasps use a pebble to tamp down the earth over their nests in the ground.

The chief conclusion is that human origins were in Africa about the time that the Pliocene became the Pleistocene—about 2 million years ago. Fossils from this period of change include at least two species of *Australopithecus*, both perhaps ranging through much of Asia as well, and several species of *Homo* that antedate modern man. Probably *Homo sapiens* displaced the others and became the sole survivor during the interglacial years between 150,000 and 50,000 years ago.

As ever more camp sites and other evidences of human presence during Pleistocene times come to light and are dated by reference to the natural decay of radioactive isotopes, it may be possible to learn the principal routes our ancestors took as they emigrated. Carrying crude pebble tools that still earlier men had invented and used for hundreds of thousands of years, these members of our species reacted as best they could to the unusual weather pattern that gripped their world. They circuited the great ice expanses in the Northern Hemisphere, and found new challenges to life at equatorial latitudes and in the Southern Hemisphere as well. No doubt many populations left no survivors. Others became separated and evolved into genetic strains or races. But the drift in the frequency of genes in each isolated population never reached a stage of true fixation before climatic changes allowed the peoples to meet and share their heritage. Mankind remained one species, variable in the extreme.

It is easy to conclude that one population became isolated in Asia north of the huge glacier capping the Himalayas, on territory where winters were long and severe but too dry to accumulate an ice cover. Among the members of this group natural selection favored features that retained heat in the human body. Following Allen's rule, the heat-producing body itself became disproportionately large in relation to the length of legs, arms, and noses. High cheek bones and pads of fat around the eyes (particularly in an epicanthic fold that makes the eyes seem slanted when open) gave these important organs additional protection. Together the pattern that evolved became recognizable as marking a mongoloid group.

In the tropical lands of Africa and Asia, tolerance for intense sunlight and needs for ease in dissipating body heat remained physiological problems. Responding through an accumulation of beneficial mutations, the populations there developed longer arms and legs and more slender build, with proportionately more surface for heat loss, and a skin so dense with black melanin granules that neither ultraviolet nor infrared from sunlight could penetrate to damage tissues. A negroid group evolved.

South of the Himalayas in Asia and across North Africa and southern Europe, the hot summers and cold winters became influential environmental factors. In response, the human inhabitants became the hairiest of mankind, with long noses serving to warm inhaled air in winter and to humidify dry air during prolonged summer droughts. Skin pigmentation ranged from very pale in the north of Europe and among the hairy Ainu people in Japan, to very dark brown in Arabia and the Indian peninsula. The caucasoids retained these features for millennia.

Subdivisions in these racial groups matched different rates of cultural development and unlike routes in colonizing new lands on earth. By the end of the Ice Age, when some of the surviving types of *Homo sapiens* began deliberately to cultivate grasses for grains and to domesticate animals, immense glaciers still diverted the pioneers away from the whole area of Hudson Bay, of the Canadian Rocky Mountains, of parts of the Kamchatka Peninsula, of mountains to the north of Mongolia, of Scandinavia, and of almost the whole of Switzerland. This left a route by which the more primitive type of mongoloids spread from China across the Bering land bridge into Alaska, east to the Mackenzie River drainage, and south over the Great Plains, to become Amerindians. With a loose social organization and few tools at that time, but with the domestic dog as a valued companion, they colonized southward through Central America, to the major islands of the West Indies, and around South America, except for its extreme tip and central heartland.

A more advanced group of mongoloid

people developed new cultural traditions in far-northern Asia. Some of them found ways to live close to the Arctic Ocean, eating mostly seals in winter and wild reindeer in summer. Others domesticated the reindeer and additional animals, including, perhaps, the horse.

At this time, around 10,000 years before the present, diminutive Pygmy people claimed most of the tropical rain forest in the broad Congo basin, and perfected a nomadic life within its shadows. The rest of Africa was divided almost equally between the Negro people to the north (from the Atlantic to the Red Sea, but south of the caucasoids in North Africa and Ethiopia) and the Bushmen, whose territory extended from the Red Sea to the Cape of Good Hope. Additional Pygmies lived in hazardous isolation in northern India and Southeast Asia, including Malaya, Sumatra, western Borneo, the Philippines, and western New Guinea. Australian aborigines and their near kin, the Veddahs, appear to have held parts of India, Southeast Asia, eastern New Guinea, and Tasmania, as well as all of the Australian continent. Like the Amerindians, they had a dog, the dingo, as a semidomesticated companion.

Already the human species differed ecologically from the nonhuman world of life in several cultural ways. Presumably language of various types was in use, allowing a transfer of information and advice, letting past experiences of the group give guidance to the young. Perhaps for the first time among primates, postreproductive individuals could influence the behavior of younger members of a tribe in ways toward actions with survival value.

In northern lands, at high elevations, and along the fringes of deserts where nights were cool, the wearing of animal pelts assembled into rough clothing became important. But the most significant cultural tool was the use of fire to warm a cave, to make edible and digestible an increasing assortment of foods, and to alter the environment directly. By setting fires in dry grassland, men learned to push back the boundary between prairie or savanna and actual forest, into regions with slightly greater rainfall and correspondingly higher productivity. Temporarily, the burning lessened also the danger of attack by grassland cats (which need cover to stalk human prey) and parasitic arachnids such as ticks and the mites known as chiggers (which perch on grass until they can catch on a passing host). An annual burn or two increased somewhat the territory in which men could hunt the larger grazing mammals.

At the end of the Old Stone Age 12,000 to 10,000 years ago, men still relied upon hunting for food that could be pounced upon, killed with a simple weapon, pulled down or dug up. Every able-bodied member of each tribal group, from the very young to the oldest of both sexes, cooperated. On this basis, about two square miles of fertile territory were required to support each person. On land of less fertility, the area to be gleaned for food had to be larger. If all of the territory occupied by people of all kinds prior to the origin of agriculture is included, it is equivalent to less than 20 million fertile square miles. If the human population had reached the carrying capacity of its environment, the maximum number was about 10 million people. Half this amount is a more probable number.

The Advent of Food-Raising
The wrinkled landscape of Asia Minor must have offered early man an extraordinary

assortment of plants and animals that were suitable for domestication. No other part of the world has provided so much or been so suitable as the site of the first civilizations. It gave us barley, wheat, oats, and rye as cereal grasses. The goat, camel, donkey, perhaps also sheep, pig, horse, dog, and cat seemingly originated in this area. Migrant mallard ducks and graylag geese may have been caught there on their wintering grounds and confined until they reproduced as a tame population.

At Qalat Jarmo in northeastern Iraq, remains of food (including traces of cultivated barley, wheat, goat, and dog) have been unearthed and assigned by radiocarbon methods a date of 6750 B.C. The older and lower levels of this archeological site yield flint sickle blades and milling stones, proving that progress had been made in handling grains. The highest, most recent levels include fragments of the first known pottery. Almost certainly the jars containing stored products of the new agriculture attracted a local mouse to move in with mankind and become the house mouse. Today none of these mice are known in the wild, but close relatives eke out a living in the same general region. Plagued with house mice, the people may have imported house cats from the Nile Valley, where these animals were revered. Cattle may have been obtained from the same source. Or the ancestral bovine may have been native to Asia Minor as well, particularly the region between the Tigris and Euphrates rivers, known as Mesopotamia, "between the rivers".

Egyptian culture, which spread along the Nile and was sustained by crops fertilized and irrigated by the annual flooding of the river banks, dates from about 4500 B.C. It took some 2000 years to reach the stage of erecting great pyramids to honor its ruling monarchs. On the island of Crete, the Minoan civilization seemingly developed as a parallel, with frequent intercommunication, between 4000 and 1000 B.C. It overlapped the Mesopotamian empires of Assyria (with Nineveh the capital) and Babylonia, the Aegean civilizations on the Greek islands and mainland, the Trojans and the Phoenicians and the Mycenaeans that all preceded classical Greece and Persia and Rome.

Systematic cultivation of cereal grains, fruit trees, and domesticated animals by the caucasoid peoples, and, to a lesser extent, by the Negro group to the south in Africa, and, after the third millennium B.C. among the more advanced mongoloid peoples in the Far East, gave an ecological advantage that by 1000 A.D. led to marked shifts in the distribution of human races. The Pygmies, who never took up agriculture or animal husbandry, diminished into four small enclaves in tropical central Africa, two in eastern India, one each in the Philippines, Malaya, western Sumatra, and western New Guinea. The Bushmen similarly shunned the new ways that evolved, and shrank into southern Africa below a frontier from approximately Angola to Mozambique. Negro people moved south, taking over the territory of Pygmy and Bushmen and occupying Madagascar. Negroid pioneers travelled to New Guinea, displaced the Australian aborigines there almost completely, then spread onward to colonize those islands in the western Pacific now known as Melanesia.

The caucasoids spread from England throughout the British Isles, over Scandinavia and to Iceland, and southward from North Africa into the expanding deserts as Arabic, Hamitic, and Semitic people displacing negroids. By contrast, in southern Japan, the white-skinned Ainus lost territory to the encroaching mongoloids of newer type. These vigorous people had enlarged their range to the coasts of the Arctic ocean as far west as Finland, and eastward across North America (as Eskimos) to the shores of Greenland, as well as southward into the East Indies and the Philippines. As Polynesians, some of them colonized remote islands in the western Pacific (Polynesia) and reached New Zealand, where they called themselves Maoris. The older type of mongoloid people disappeared completely in the Old World, but survived without interference from outside while occupying the rest of the Americas—still as Amerindians.

Shifting populations of the various racial groups encountered many unfamiliar ecological situations. Yet so long as domesticated plants and animals could be substituted for the endemic vegetation and animal life, people could concentrate on improving their increasingly artificial culture, virtually independent of geographical location. For this reason, the cultivated kinds of plants and animals have been credited with generating culture and have been given the name cultigens. To possess them did not free anyone from attack by local predators and parasites. Indeed, the cultigens constituted a form of wealth that was vulnerable to plant diseases, pestiferous animals, and nomadic people. Yet with cultigens to harvest, men had no need to spend time in hunting or gathering wild foods. They could learn, instead, to tolerate neighbors in urban communities and to benefit through division of labor. Meanwhile, the cultigens generally lost the ability to survive without human care. Mankind and cultigens became completely mutualistic.

It is strange how few kinds of plants and animals were domesticated freshly in parts of the world other than Asia Minor. The caucasoid people of India possessed many of the cultigens of their forebears who, around 2500 B.C., came from the northwest in waves, Aryans following Dravidians, who came after still earlier pioneers. They knew cereals but not rice and found a different type of cattle (the hump-backed Brahmin). They did domesticate the water buffalo, learn to tame and employ the native elephants, and began breeding a fowl in captivity—a small bird yielding eggs to people who had a religious aversion to meat.

The mongoloid people to the north and east in Asia benefitted from cultigens adopted in the Chinese culture, which evolved in almost complete isolation. It was based upon a wetland grass—rice—along with soy beans, egg plant, bamboo shoots, and water chestnuts. The widespread mallard ducks and a distinctive goose called the swan goose came into northern China each winter, and there were caught and developed into captive domestic forms. The equally widespread Eurasian hog became the China pig, as a counterpart to its European forms, such as the Duroc and the Polish. These Chinese cultigens were shared to some extent through the mountain passes with the people in the Indian sub-continent. Along the boundary, the Himalayan yak became a source of dairy products, hides, and meat, as well as a beast of burden. But it could not stand environmental conditions at low elevations.

Curiously, the fowl and the pig were carried along through the East Indies into the islands of the western Pacific by people with quite different beliefs. The islanders discovered how to make edible the starchy underground stems of taro, as a low-protein food known also as dasheen and poi. Yet these travelers depended upon the prevalence of coconut palms, which presumably had colonized all suitable islands, however remote, long before people did. Coconut palms provide not only building materials, fuel, and food, but also buoyant fruits containing uncontaminated liquid (mostly water) in abundance. With a few dozen coconuts in an outrigger canoe, an islander could start off on a voyage of discovery, his thirst provided for in unsinkable containers.

In tropical America the mongoloid people (Amerindians) had no widespread meat animal, unless it was the native turkey, until Spanish colonists brought some from the Old World in the 1500's. The Andean people had domesticated the mountain camels, and developed distinctive llamas and alpacas for many uses. But these cultigens could not be shared through the lowlands because, much like the yak, they were specialized for life at high elevations. All Amerindian cultures were built on maize and beans (kidney and lima), with additions between the tropics of manioc (or cassava) in the lowlands and small potatoes in the high Andes. Throughout the warm Americas, the Amerindians were delighted to add rice to their beans and maize products. But few have yet developed an appreciation for meat.

In Africa, the negroid people had millet and sorghum and, in the warm areas, breadfruit trees. They were glad to get maize and potatoes after America was discovered. In varying degrees they accepted cattle, sheep, goats, and the horse as benefits from the outside world. Over much of their territory they could not raise these animal cultigens because an endemic disease of wildlife proved fatal when the parasitic protozoans (trypanosomes) were shared by the blood-sucking tsetse flies.

Zoonoses, local predators, pest insects, and fungus diseases all became problems of increasing magnitude as cultigens were introduced wherever man settled. Occasionally the native animal merely benefitted from the disturbed ecosystem adjacent to farms and habitations. Far more native robins and introduced starlings live close to man rather than in the open country. In the Americas, the most widespread of native cats—the cougar or mountain lion or puma—responded quickly to the introduction of horses and cattle by pouncing on these less wary domesticated beasts instead of deer. Man responded by declaring the big cat an outlaw and trying in every way to eliminate it—with striking success. In Colorado, a black-striped yellow leaf beetle (*Leptinotarsa decimlineata*) transferred its attention from native members of the nightshade family to the cultivated potato plants that pioneers set out. It became the detested potato beetle, and spread eastward along the supply line, eventually to Europe from which the "Irish" (but truly South American) potato was introduced to North America.

The advent of agriculture, which began about 12,000 years ago, led briefly in most places to an increase in productivity and in diversity of the ecosystems. Native herbivores and carnivores had new resources to test and possibly to use. Successional stages were set back from climax conditions and started in ways tailored to high yield. The small farm and the small herd of domesticated animals added to the variety in the biota. So long as they remained small, the continuous forest in which the farm was but a clearing, or the extensive

grassland where the herd wandered, remained ready to reclaim the small acreage if it was abandoned. But as soon as man turned toward monoculture and claimed large areas for the exclusive use of his cultigens, diversity fell sharply. Total productivity dropped too, although yield of the one crop might be high and rewarding.

The Man-Altered Landscape

With cultigens to rely upon, most men could socialize and specialize in urban life. Historians began to record the doings of their leaders, the dates of battles, the names of generals, the rise and fall of capital cities and of empires. Rarely did they document the accompanying deforestation of watersheds, the lessened flow of essential rivers in late summer, the failure of crops through drought, the erosion of soil, the lengthening of supply lines by road and aqueduct and ship, the resistance to increased taxation and acts of sabotage as military might was turned to safeguarding the flow of food and other natural wealth from conquered lands along the fringes of each empire. These changes, which can be appreciated today, were the ecological consequences of a dependence on cultigens. Barely noticed, the landscape became exposed to the sun and drier, bringing each civilization in turn to a point of crisis. The final battle provided the coup de grâce for an enterprise already dying, with the participants generally unaware of the true causes.

Often the soil eroded because of intensive, desperate farming and the activities of sheep or goats. Particularly in the mountains, erosion began as soon as the forests were felled. It continued until, as in so many Mediterranean countries and the Arabian peninsula, little that men find valuable will grow. This is the message that can be read

Reforestation
Eleven years after this Missouri farm was set out with young slash pines (top), a respectable forest of trees had developed (bottom), which will yield a profitable harvest when 20 to 25 years old. During the growing period, the forest improves the water-holding capacity of the soil and provides a habitat for many woodland animals and plants.

from the Biblical account in I *Kings*, v, telling of an international agreement under which Solomon sent "fourscore thousand hewers" to cut giant cedars from the hills of Lebanon, for construction and ornamentation of his temple and palace. Today it is hard to imagine where he could find 80,000 skilled woodcutters to attempt such a task, or that the Lebanese mountain slopes could provide the trees he wanted. The mountains there have been cleared of all except one small protected grove of cedars at Las Cedres. To see trees of this species on native soil in their former glory, a person must now travel to remote parts of Turkey, where the last forests of them are being felled.

The fall of Rome in 476 A.D. is often used as the starting point for a millennium known vaguely as the Middle Ages. It is a Western concept, for no matching changes in human culture affected the isolated people in the Orient. Formerly called the Dark Ages, it was a period of religious and feudal strivings after power, most famous for its kings (such as Charlemagne), its popes, and the various crusades. The crusades between 1070 and 1296 did increase ecological awareness, for observant men went on those expeditions. They saw that the plants and animals of the Holy Land were not those of northwestern Europe. A few travelers who were versed in the writings of Theophrastus and Aristotle came to realize that the revered books applied well to the living communities in Asia Minor, but that new studies of nature were needed elsewhere. Another ecological consequence of the crusades was the introduction into Europe of the house mouse and the black rat (*Rattus rattus*). The rat, which probably reached Asia Minor originally from northwestern India with the armies of Alexander the Great, climbs readily. It became the roof rat in thatched houses, descending in darkness to feast on human food. Its fleas spread temporarily to feed on human blood, and transferred the parasitic bacteria of bubonic plague from rats to mankind. This devastating disease, known also as the "Black Death," can be spread among urban populations by the body and head lice of man. The rickettsias of typhus fever follow the same route.

The Middle Ages saw one famous attempt at communication between the mysterious East and the West. In 1295, Marco Polo returned to Venice after a prolonged stay in Mongolia at the court of Kublai Khan. But his fellow Venetians proved unwilling to believe the existence of so sumptuous and advanced an Oriental culture. Two centuries passed before sponsored expeditions of exploration began returning with news of distant ways of life, bringing samples of unfamiliar plants and animals as curiosities or potential cultigens.

Subsidies of Power for Man

Historians rarely emphasize that the cultural art of smelting iron, which began in Solomon's time (about 900 B.C.), had ecological consequences beyond making iron available for swords and plowshares. Certainly a farmer could be much more efficient in raising crops when he could substitute an iron plowshare for a plowing stick. He could use more land for farms and pastures where forests previously grew. His farm products allowed an increase in urban populations, facilitating the social interchange necessary for further improvements in technology.

As it grew and spread, the iron industry made the greatest single demand on wood as fuel for the smelting operations. De-forestation progressed over Arabia and most of Europe until, in the late 1600's, these areas could no longer supply the quantities of wood needed. A shortage of lumber, more than anything else, led to use of brick and stone in the rebuilding of London after the Great Fire of 1666. But this too had an ecological effect on man. Walls of solid construction and roofs of tile and slate (necessitated by a shortage of thatch grass) offered far fewer hiding places for black rats ("roof rats"). Indirectly the use of wood for fuel in the iron industry made people safer from bubonic plague and typhus fever. Another step in the same direction came in the early 1700's, when brown rats (*Rattus norvegicus*) reached Europe on ships from the Far East. They invaded the cities, driving out most of the remaining black rats before settling down in sewers and garbage dumps, farther from people than a thatched roof.

As the demand grew for a substitute fuel in smelting iron, men turned from use of trees to coal. Instead of releasing through combustion the solar energy stored by green plants during a few preceding centuries, they began exploiting energy that had been in storage for 60 to 340 million years. Subsidized in this way with power that could be used only once, technology and culture advanced rapidly. So did alterations of the human environment, as mines to get coal were dug ever deeper into the earth and the sooty fumes from incomplete combustion increasingly polluted the urban air.

New machines, invented from about 1450 on and made largely of metal, gave Western man more power to expand and to convert environments for human uses. Prime among the tools for domination were firearms, which let determined pioneers

In dividing the land areas into agricultural and urban categories, man has provided no program of urban renewal as a counterpart of the great biogeochemical cycles that tidy up the wastes from living things. Amid the outmoded buildings in old parts of big cities, many children now grow up seeing little that has a lasting value. They are surrounded by trash and wastes that pile up more rapidly than decomposers can dispose of them. Ecologists are seeking ways to speed up the recycling of the inevitable products of mankind.

enter strange territories, subdue indigenous people, and create a facsimile of European farms and cities far from home. Even the crude matchlock muskets that became available in 1450 enabled a handful of Spanish conquistadores to destroy the Inca and Aztec empires in tropical America. The Pilgrim Fathers and later colonists to the North Atlantic coast were similarly outnumbered by the Indians, but the Indians were outgunned.

Equipped with firearms, the caucasoids extended their previous range in the world and their culture south in Africa to the 15th parallel of North Latitude and north to the 30th of South Latitude. They further constricted the territory of the Australian aborigines, and dominated the Maori people in New Zealand. They extended a wedge across southern Russia to the Sea of Japan at the expense of the more advanced mongoloid tribes. They wrested most of the Americas from the Amerindians. None of the other racial groups fared so well. The Pygmies almost disappeared, and the Bushmen too, unable to compete with members of mankind who possessed both cultigens and metal tools.

Slaves from eastern and western Equatorial Africa joined the caucasoid population involuntarily in the West Indies from 1502 on, in southeastern North America beginning in 1619, and in northeastern South America and South Africa after 1567. The ecological effects of slave-trading were numerous, among them the introduction from the Old World to the New of grain sorghum, the malaria parasites (*Plasmodium*), and both the yellow-fever mosquito (*Aedes aegypti*) and the disease it carries. Nor did the emancipation of the slaves (1833 in South Africa and all British colonies, 1863 in the United States, and 1942 in Ethiopia) do more than reshape racial interactions and the patterns of land use.

During the two centuries between 1650 and 1850, wilderness frontiers were pressed back in the Americas and Australia, disrupting ecosystems and dislodging both the indigenous people and the wildlife. Raw materials were hauled ever faster by ship, road, and rail to urban centers where increasing numbers of people labored with machines.

Engines using coal took the place of

Until three centuries ago the production and distribution of food and materials was achieved almost entirely by manpower and the work of domestic animals, supplemented by the inanimate power of the wind and of flowing water. Some of these ancient ways of life have continued where they fit a harsh environment. The indigenous tribesmen of central Australia (below) still survive around the fringes of a great desert by catching kangaroo for food and by hauling the useful parts of the carcass back to their nomadic families. These people cannot compete well with civilized man, who has taken over much of their hunting territory and converted it to farms. Man has learned to get supplementary power from fossil fuels and to use it in unwieldy machines to make and move both himself and his products, and in doing so has removed himself from being an integral part of nature (right). To him the mixture of life, of goats and cattle, in the cities in India seems incongruous (far right). Personal involvement is still essential toward inventing new ways, new products, toward assembling and monitoring, and repairing equipment that will be replaced by new kinds in a few years, and for inspecting (below, right) or packaging products. The versatility of the human mind and senses is not made obsolete by machines, but it is debatable whether it is stimulated by them.

In Turkey and parts of North Africa, dwellers in arid lands have dug their housing underground where the thermal inertia of the earth itself insulates them from the daily heat and chronic drought. This procedure, although successful for a small community, has not yet proved adaptable to modern techniques of water distribution, sewage disposal and ventilation.

windmills and water mills, which required no fossil fuels for energy and produced no pollution. In enlarging areas of Europe and North America, soft-coal soot and fumes from industial smokestacks settled on the cities and adjacent countryside. Storm sewers, which were built originally to carry off rain that fell on city roofs and pavement, were put to use carrying human and industrial wastes to the nearest river, lake or coast. Almost no one calculated the rising cost in pollution of the environment, or chose to pay anything toward preventing it. The deterioration increased faster in Western nations that in those of the East because of unequal industrial activity and the spiral rise in buying power and technologial development. The unpleasant facts were never faced, in a general belief that they were temporary, or unavoidable, or actually justifiable along the way toward gaining "control over nature."

In subtle ways, industrialization and urbanization led to changes in human survival and in cultural practices. To accommodate more factories and homes for workers, lowlands and wet places in and near cities were filled and built upon. Although wildlife found fewer opportunities, it was scarcely missed. Mosquitoes, however, lost many of their breeding places. With the decrease in vectors, malaria diminished as a scourge in urban parts of the developed countries. The reason was not obvious, for the cause of the disease had not yet been discovered. Today it is hard to think of malaria as a major disease in Boston, London, and Moscow, as it was two centuries ago. At that time more people lived longer, but at the same time, the birth rate decreased in a pattern that still has no counterpart in undeveloped countries.

Artificially Simplified Environments

Since 1850, mankind has made tremendous strides toward replacing the natural world in the vicinity of people with an artificial environment. In many countries, 80 per cent of the human population lives in cities where native plants and animals are mostly inconspicuous and insignificant in affecting man's evolution. Suburbs and countryside have changed almost as much. Highways, railroads, and airports provide for rapid transportation of industrial products from the cities, produce from agricultural areas in the opposite direction, and people between all parts of the world. No place on earth is now more than two or three days' travel from any other. In Anglo-America, the number of family-size farms and ranches, with 100 or fewer acres, decreases continually because they are less efficient as business enterprises than large, professionally-managed holdings of 1,000 acres or more; these grow in number and degree of mechanization. Wilderness shrinks. The frontiers are almost gone.

Often we regard as the mileposts in progress the technological achievements that are based upon an increasing use of non-renewable natural resources. More significant over the long term are the revolutions in scientific understanding that made possible both the present and all future gains. It is difficult to realize how high the full cost of modern culture has risen in the industrialized nations, or how disproportionate is the distribution of natural resources, both renewable and non-renewable.

We pride ourselves on being able to produce rice in the lower Mississippi region at so low a cost that the product can be shipped to the Orient and sold there below the price of local rice raised in hand-tended paddies. We admire the management of the Mississippi bottomland, which is carefully diked and ditched by machines. The ditches serve alternately for irrigation and drainage, according to the action of motor-powered water gates under the command of a man in a remote control station. From this isolated point he monitors the water level in every field by reading the dials connected to ingenious electrical sensors at each gate. When the fields are ready and the weather auspicious, the rice is sown from a low-flying airplane. Applications of fertilizer, weed-killers, insecticides, and fungicides are made periodically from the air, with no person on the ground. Only when an airborne observer reports that the crop is ripe are the fields let drain thoroughly. Under contract, men arrive with machines to harvest the grain. Before they depart, they prepare the field for the next sowing. When the cost of seed, chemicals, maintenance, operations, shipping the finished product, and the taxes are all paid, a profit remains from the sales. Yet the energy expended through use of fossil fuels to produce the machines and chemicals and to utilize them is several times as great as the food energy that the rice plants captured from sunlight and incorporated into the kernels.

Without realizing it, we subsidize many modern enterprises in food production by spending as many as five kilocalories from fossil fuels for every kilocalorie in the product as delivered to the consumer. Each loaf of bread on the shelf of the supermarket and each package of frozen vegetables in the freezer is a similar trade-off, with only a fraction of the cost in kilocalories represented by nutritional values inside the wrapper.

That a technologically sophisticated program can be established profitably in one part of the United States does not mean that it can be applied in other areas of the country or in other countries. The climate must be right. Resources, such as abundant water at low cost, are fundamental. Skilled personnel and the temporary use of expensive machines, such as aircraft and harvesting equipment, are essential. The enterprise may thrive only while it can qualify for a favorable tax treatment, and while it is not required to ensure that wastes from chemical additions taint neither the air nor the runoff water. These limitations and hidden subsidies are often overlooked in recommending that techniques be imitated elsewhere.

In parts of the world where technology is well advanced and education available to virtually everyone, medical care and public health measures are both reasonably efficient. Communicable diseases have become as rare as predators dangerous to man. Yet the ecological consequences of this simplified environment prove far more complex and far-reaching than anyone anticipated. In the United States, for example, the expectation of life at birth among people in the class that buy life-insurance policies rose between 1900 and 1965 from 47.3 to 70.5 years. More than 23 years were added to the lifespan of the ordinary inhabitant, because more babies survive and people stay healthy through childhood into normal old age. Yet these 23 extra years increase the total demand for space, for drinkable water, and adequate food, for all the material things that go with modern civilization, and for recreational opportunities in proportion to affluence.

Citizens in the undeveloped countries see that medical expertise and medicines can be imported. But the saving in lives and extension of life upsets the previous relationship between natality and mortality, leading to a sudden increase in population. Leaders in the undeveloped nations are asked for miracles to meet the extra needs for food, gainful employment, and services such as education. Yet each avenue seems blocked by some scarcity of a natural resource. The uneducated people can scarcely be expected to understand and accept the limitations. The educated leaders see little hope through facing the facts.

The geographical distribution of important resources is due to events in the remote past. Swamp forests of the Carboniferous era left coal in a few countries but not in all. Microbial plants in shallow seas of the Mesozoic may have provided the crude oil and natural gas where they are found today. The glaciers of the Ice Age left a legacy of freshwater lakes and streams in parts of northern Europe, in Canada, and the northern United States. Igneous activity determined the erratic distribution of iron ore. Yet technology could develop and culture leap ahead where coal and iron ore occurred in proximity, and where water was naturally abundant. Men recognized the opportunities and seized them.

The United States found itself blessed with more than half of the coal and lignite that could be exploited by modern methods, and almost half of the iron ore that contains 20 per cent or higher concentrations of iron. Along the boundary shared with Canada, the Great Lakes hold about a fourth of the world's fresh water. Large additional amounts are available to agriculture, industry, and municipalities in the basins of the Mississippi and other rivers. (Another fourth of the total is in the Amazon River, where Brazilians use it chiefly for navigation; the rest of the world must get along with less than half of the earth's fresh water.) The geographical distribution of crude oil and natural gas are important now that these fuels so greatly augment the use of coal for energy. So far as is known, the United States has about a fifth of the world's crude oil and almost half of the natural gas. With only 6 per cent of the land and 5.5 per cent of the world's population, this one nation has exploited these resources for two centuries and forged ahead.

The gain within the artificial environment that men created for themselves in Europe and Anglo-America has enabled the developed nations to buy additional resources from world markets. The United States is a valued customer for raw materials from the less developed countries. Currently it consumes from local and foreign sources about 50 per cent of the newsprint and the synthetic rubber, 38 per cent of the tin (virtually all imported), 37 per cent of the fossil fuels, more than 25 per cent of the steel, phosphate rock, potash, and nitrogenous fertilizers, and perhaps 30 per cent of all renewable natural resources. The balance in each instance is divided in extremely unequal portions among the other nations.

The total production of commodities in the world can also be analyzed, as H. R. Hulett has done recently, in terms of the number of people who could be supported at the average standard of living in the United States: on food, 1,300 million (not the 3,700 million alive in 1970); on forest products, 1,000 million; on fertilizer, 900 million; on steel, 600 million; on energy, 500 million; on aluminum, 400 million; and so on.

In the developed countries, the artificial environment has been expanded as though fossil fuels and other natural resources were in limitless supply, and man was somehow exempt from the ecological limitations seen among other living things. In no advanced country today is the human economy in step with the capture of solar energy through photosynthesis or the great biogeochemical cycles that renew the supplies of fresh water and other chemical nutrients. We forget that a mere three centuries ago this extravagant exploitation of non-renewable resources got under way. We prefer to believe that, while technology grows increasingly wasteful of energy in extracting lower grade ores and moving crude oil or fresh water for ever greater distances, some miracle will save our standard of living and our environment.

Our present way of life has been called a "through-put" system, in which raw materials are combined into products for temporary use and then rejected as trash. Little that is lasting accumulates. Yet the activities required for production all release heat and other byproducts with an ecological impact. Many of them can no longer be ignored. Enough damage has already been done to the ecosystems on land, in fresh waters, and throughout the seas that it now seems unlikely that a total human population of 500 million could be sustained after the year 2000. To continue operation, it would have to manage with less expenditure of energy, and with an utterly altered economy geared to recycling of wastes.

The technological solutions to modern problems generally shift the conflict of interests into different areas, making the problem bigger. A small disaster in the present is averted by virtually ensuring a major catastrophe in the near future.

The long-term effects of engineering applications in a developed and an undeveloped nation can be compared in the management programs for the Colorado River and the Nile to meet specific needs. Both rivers rise in snow fields on high mountains, and flow through deserts on their way to the sea. The Colorado is the most manipulated large river on earth. Before it has gone 100 miles from its source on the western slopes of the Rocky Mountains, a major fraction is diverted through a long tunnel to the eastern slopes, to irrigate farm land and to supply municipalities in the rain shadow of the range. While the remainder of the river, with additions from further tributaries, proceeds through a corner of Utah and crosses Arizona, its potential energy is tapped for hydroelectric power at one great dam after another. The volume released at each stage is adjusted carefully according to a master plan. One huge portion must always be available for diversion to Los Angeles and other cities along the Pacific coast as far south as San Diego. Farmers in the Imperial Valley of California must be able to buy cheaply the irrigation water they need. Arizona farmers get a share and want more. So do Mexican farmers just south of the border. Their waste water is returned to the United States. It joins the runoff from the Imperial Valley, and flows to the Salton Sea, where it evaporates. The Colorado River no longer has a mouth. Much of its water emerges from the sewers of Los Angeles, but the effect on fisheries offshore has not been calculated.

For millennia the Nile has been a managed river too. Farmers along its flood plain in northern Egypt have labored to get the greatest benefit from the annual spreading and withdrawal of the waters, trying to slow the drying of the desert edge until their crops were ripe. Egyptians dreamed of building a huge dam at Aswan, to expand their irrigated land and to produce hydroelectric power with which nitrogenous fertilizer could be manufactured. Lacking the resources for so massive an engineering feat, the Egyptian leaders sought foreign aid and received it from the U.S.S.R. The Aswan dam is to be finished by 1973. But dreams of having an abundance of food for Egyptians and a surplus for export have faded. In the years since the foreign aid began, the population of Egypt has increased so rapidly that by 1973 the extra million acres of irrigated land will add only about 1/10 acre per person—a scarcely perceptible gain where people are already undernourished. Yet already the side effects of the new dam are causing consternation. The dread disease of blood flukes (schistosomes or "bilharzias"), whose vector is a snail in irrigated fields has spread throughout the expanded agricultural land in southern Egypt where it was unknown before. In northern Egypt, crop yields are falling off because the Nile no longer brings an annual load of free, fertilizing silt. Instead, the silt settles behind the dam, diminishing the storage capacity of the lake.

So much moisture evaporates from the lake and the irrigated fields it supplies that less water reaches the delta country at the mouth of the Nile; salination of the soil there is now a problem. Even the water released for agriculture below the dam is significantly saltier and less suitable for crops. And in the eastern Mediterranean Sea, the sardine fishery upon which many people depend for a livelihood and food, is falling off rapidly because the Nile no longer contributes dissolved nutrients to the drifting green plants (phytoplankton), which form the producer level of the ecosystem.

The managed water of the Colorado and the Nile now support larger populations of people than could otherwise live in areas that are naturally arid. The artificial environments are simpler, with fewer kinds of plants and animals. Those that share the areas with man are not the native species, which possess extraordinary adaptations toward survival in a climate with scanty rain. Instead, they are cultigens and weeds. They flourish only so long as man has energy to spend in maintaining his artifice.

10 Ecological awareness

Among the scientific understandings that are relevant to man's future in a man-altered world, half a dozen stand out. Each of them originated at a technical level, but have spread into the realm of general knowledge.

1. The theory of organic evolution, supported by evidence first by Charles Darwin (1858 on) and broadened now to apply to all living things from the origin of life to the present. Today, new combinations of selective forces in the environment affect mankind among the animals, the plants, and the microbes—all products of past environments. New species replace old ones by competitive exclusion.

2. The proof by Louis Pasteur (1864) that in the modern environment even microbes arise only from pre-existing life (biogenesis). The possibility that, on the primitive earth under an atmosphere free of oxygen, abiogenesis could have produced the first organisms was proposed by A. I. Oparin in 1938. The suggestion led to laboratory testing, which showed just such abiogenic production of amino acids (1953), adenosine triphosphate (ATP, 1962), and other essential organic compounds now synthesized only by living cells. Physical laws of chance and of thermodynamics were seen to suffice for life to arise spontaneously anywhere in the cosmos when knowable conditions favored the necessary combinations.

3. The germ theory of disease, as offered by Robert Koch (1876) after he discovered that a specific microbe causes the disease known as anthrax. The ecological discoveries of phagocytosis (Eli Metchnikov, 1884), of vectors of disease organisms (mosquitoes and malaria by Sir Ronald Ross, 1897), of antibiosis (Sir Alexander Fleming, 1929), and of pathogenic organisms resistant to antibiotics (1958), all led to important and effective changes in medicine and public health—extending the average human lifespan.

4. The rediscovery of Gregor Mendel's laws of simple inheritance (1900). This was followed quickly by the chromosome theory of heredity (Walter Sutton, 1902), the theory of the gene (Thomas H. Morgan, 1916), more recently by the model of any gene as a molecule of deoxyribonucleic acid (DNA, 1953), and the subsequent decoding of genetic information. All are extensions of a single concept that applies equally to microbes, plants, animals, and mankind.

5. The recognition of a virus as an autocatalytic disease agent of chemical simplicity (Wendell H. Stanley, 1936), bridging the assumed gap between the nonliving world of pure substances and the world of complex life. Subsequent studies of viruses revealed their involvement with the genes of the host organism, and led to new knowledge on protein synthesis, enzyme action, and molecular biology.

6. The realization that explanations at the molecular level are too simplistic. They cannot account for the versatile interactions among whole organisms, such as a lion stalking an antelope that is eating green plants which carry on photosynthesis. Nor is it enough to know that all living things are transferring and losing energy captured from a thermonuclear reaction in the sun. A fuller comprehension of ecological complexities, mostly since the middle 1960's, has tended to unify biology, geology, economics, and sociology toward admitting man's dependence upon a nonhuman environment. The price of oversimplifying or polluting that environment could be extinction for all life on earth.

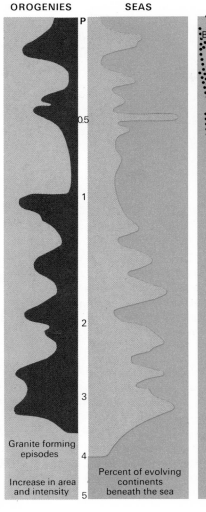

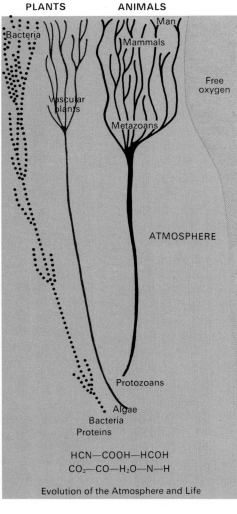

OROGENIES	SEAS	PLANTS	ANIMALS

Bacteria

Vascular plants

Man

Mammals

Metazoans

Free oxygen

ATMOSPHERE

Protozoans

Algae
Bacteria
Proteins

$HCN-COOH-HCOH$
CO_2-CO-H_2O-N-H

Granite forming episodes

Increase in area and intensity

Percent of evolving continents beneath the sea

Evolution of the Atmosphere and Life

Plotted on a logarithmic scale measured in billions of years, the upward evolution of the planet Earth, its epicontinental seas (center), its life and atmosphere (right) including free oxygen, and its major episodes of granite-mountain formation (left) can be seen in a long perspective. The 0.5 level— half a billion years before the present—is in Cambrian times. No one knows how much more than 5 billion years ago the planet formed, or how long afterwards water began to fill the ocean basins, and then to overflow the low parts of the continents.

Ecological awareness has caused men to reassess the world. Among the first to attempt such an undertaking was the distinguished French chemist Antoine L. Lavoisier, who tried to analyze the atmosphere and the ocean. Between that time and the middle of the twentieth century, pollution of the atmosphere seemed local and no significant changes of any kind were suspected in the constitution of the seas. But since World War II, contaminants have become global. Today their concentrations are changing so fast that the most important feature of a detailed chemical analysis is the date at which it is made. No longer can the biogeochemical cycles dispose of the wastes thrown off by civilization.

That wastes should now take on this new significance is paradoxical, for living systems operate efficiently in the face of waste that man regards as incredibly extravagant. Often more than 99 per cent of the spores and seedlings of plants are wasted. A comparable mortality is evident among the eggs and young of oyster, cod, sparrow or mouse. This waste of energy and materials and potential by the members of the species that start the fresh lives proves to be essential: it keeps wild populations fluctuating close to the carrying capacity of their environment. The complex natural world is well organized to dispose of the energy and recycle the raw materials at the rates that have evolved in the past.

No natural process has ever continued to "inject" wastes into the atmosphere at the rates that now are characteristic of civilization. The largest annual tonnage is gaseous, including carbon monoxide and dioxide, oxides of sulfur and nitrogen, and hydrocarbons. Local accumulations of these materials gave the earliest broad symptom that man was misusing the environment. It came, appropriately, as smog over an especially artificial city (Los Angeles) as a phenomenon of the 1940's. Currently the United States contribution to injection of these gaseous wastes into the atmosphere accounts for from a sixth to a third of the totals for the world. Half of the man-made dust comes from the United States. Solid matter and smoke particles from use of fossil fuels and manufacturing processes now equal in tons per year the natural formation of dust from winds over dry land and belching volcanoes.

Lead, which has been known for many years to be a dangerous poison, is injected into the atmosphere from vehicles that burn gasoline fortified with tetraethyl lead. During 1968 in the United States, automobiles produced about 350,000 tons of these pollution particles and aircraft another 25,000 to 30,000 tons. (The totals for the whole Northern Hemisphere were probably several times as great.) Although winds can carry lead particles for considerable distances, about 100,000 tons of lead particles of American origin were believed to fall on American soil during the same year. Rivers from the continent discharge about 150,000 tons of lead annually into the oceans at their mouths. But ecologists credit at least half of this river-borne lead with originating in natural erosion of lead-bearing rocks. Oceanographers accept this estimate, and suspect that it corresponds closely to the

The gannet (above) and the guillemot (above, right) are doomed to slow death by the floating oil that coats their feathers. Oil spills in the North Atlantic Ocean contaminated the water in which these birds dive for food. Their death is part of the price of using petroleum as fuel in the current pattern of industrial civilization. Poisoned rabbits on an Australian grassland (right) are a grim reminder of the tragic consequences of deliberately introducing exotic species. In this instance, European rabbits were set loose as game for English sportsmen who found little interest in shooting the native marsupials. The rabbits prospered and soon competed seriously with man's introduced livestock. Now that myxomatosis, a disease from South America, has decimated the European rabbits on Australian soil, there is a vigorous campaign to eliminate the remaining rabbits with poisons. The danger is acute that the poisons will destroy instead the predators that are now also assisting in the control.

natural rate at which the oceans can get rid of the lead they receive by incorporating it in sedimentary deposits. Yet if nearly 75,000 tons of additional lead are coming from the rivers each year now, lead is being stored on the soil and the rate of lead discharge can be expected to rise. Furthermore, the oceans are receiving annually from America about 250,000 tons of lead as dust. No one can guess what effect this will have, how much the concentration of the poison will build up in the oceans because of lead pollution that has already taken place, or what can be done about it.

A large proportion of all wastes eventually find their way to the oceans. No one knows yet how to measure the delay or "residence time" for pollutants that fall on land. Those that reach a river are flushed out in a year or less. Those in lakes may be stored for as much as a century. In the oceans, however, a balance should eventually be struck between any constant rate of addition and some rate of sedimentary depositing. Most of the durations of storage in the ocean waters prove to be more than a century. In some cases they may be millions of years.

At present the concentration of certain organic poisons is known to be building up insidiously in soils, aquatic environments, and the bodies of living things. The inertness of these materials extends to decomposition too, for few decomposers produce enzymes that will simplify the synthetic molecules. Most noted of these materials are the chlorinated (or halogenated) hydrocarbons, such as dichloro-diphenyl-trichloro-ethane (DDT) and its equally toxic products (DDD and DDE) of slow, spontaneous, partial decomposition. Since World War II, vast quantities of these materials have been applied to forests and farm lands, either as dusts or in sprays with an oil base. Almost as dangerous may be the polychlorinated biphenyls (PCB's) that have been discharged into streams in wastes from industry. The rate at which decomposers render these materials harmless is so slow that at least half of all the chlorinated hydrocarbons used to date are believed still present in active form.

Sensitive tests that have been invented recently reveal that DDT and its relatives are now distributed from pole to pole, carried by winds and ocean currents far from the sites of application. The average concentration of DDT in sea water is now greater than in fresh water. No reliable measurement for the average amounts in soils is yet available. Yet ecologists feel confident that so much is in temporary storage on land that the toxic residues reaching the oceans will increase the concentration there for many years. The peak might come about 1985 if no more of these materials were applied anywhere on earth. Less is being used in some regions and on pests that have become immune to the pesticides. Laws have been passed in many states and nations forbidding most applica-

tions of the chlorinated hydrocarbons (except in emergency!). But the manufacture and sale of these materials continues, for sale to the undeveloped nations—as though *where* the poison is injected into the ecosphere makes much difference.

DDT and its related compounds penetrate the integument of any insect that comes in contact with them. Yet these materials have been applied as though they were fertilizer. By using a surplus because they are cheap, blemish-free fruit is often obtained. But fully half of the pesticide misses the target animals. It blows away as dust, or washes into the nearest stream. It destroys natural predators and parasites, which would otherwise help control the pest, and kills harmless and beneficial organisms alike, upsetting many food webs unrelated to the pest. None of these ecological costs are charged back to the user of the pesticide.

Persistent poisons have led to the phenomenon of biological magnification, whereby the concentration of the toxic substance increases progressively to the highest trophic levels. The effect was noticed first in robins that kept eating earthworms where each worm had accumulated some DDT from decaying elm leaves on which the pesticide had been sprayed. When enough DDT reached the robin by this moderately direct route, reproduction ceased even if the adult bird itself did not die.

In the oceans, the chlorinated hydrocarbons are absorbed by diatoms among the drifting plants near the surface. Being soluble in lipids and not in water, the poison is transferred into the oil droplets within the diatoms and stored there; the amount may be trivial, perhaps only 0.01 parts of DDT per million. Small animals in the surface layers of the ocean eat the diatoms, and are eaten in turn by copepod crustaceans the size of a grain of rice. At each step the oil or its products of digestion go into animal fat reserves, without either altering the pesticide or getting rid of it through excretion. The copepods soon have an average concentration of DDT of 0.04 ppm. Small fishes that eat the crustaceans accumulate the chlorinated hydrocarbons until they contain about 0.23 ppm. Larger fishes, which eat the smaller ones, are commonly in the range from 1.2 to 2.0 ppm. Fish-eating birds, such as terns and herons, may show from 3.2 to 9.6. Large birds of prey and scavengers, such as the bald eagle, sometimes contain 10 to 200 times as much. Age-old feeding habits lead these birds to devour any unwary fish or other prey animal that is behaving abnormally; in this way the eagles tend to select the fishes with the most poison. Soon the bird has so much in its body fat and includes so much in the yolk of each egg that its reproduction falters. Population numbers of the American eagle, the peregrine, the pelicans and many other birds have plummeted recently, as the concentration of DDT in their fat and egg yolk has risen. Experimental tests indicate that no other cause for their failure to reproduce need be sought since the measured levels of poison are consistently effective.

Even human body fat now shows a load of DDT and its derivatives. By 1964 in the vicinity of Delhi (India) it reached 26.0 ppm, while in the United States an average amount was 7.0 ppm. Among American women, the concentration of these poisons in breast milk exceeded the amount allowed by Federal regulations for interstate transport of commercial foods in saleable containers.

As ecological awareness grows, so does

a feeling of futility. The goals of civilization seem incompatible. On one hand, we try to reduce pollution in the forms of dust, gases, soluble substances, slow-decomposing materials, radioactivity, and heat. Our aim is to get these wastes from mankind down to the level at which the great biogeo-chemical cycles could dispose of them with no dangerous accumulation. To stop pol-lution altogether is clearly impractical and unnecessary. On another hand, we try to prepare for a continuously increasing human population, and give no really serious thought to either the meaning of the slogan Zero Population Growth or the more drastic reduction that would leave only 1,000 million or 500 million people on earth. We approve of the aspirations of the poor to possess and enjoy the standard of living of the moderately rich in the United States today, rather than a deliberate lowering of the standard so as to require less energy, less nonrenewable resources (or just what can be recycled), and to live on the products of soil and sea to the limit that can be dreamed of as "perpetual yield."

Our chemists invent substitutes for wood, since it is scarce; they produce plastics that are not biodegradable. We make waste plastics disappear by incinerating them, using fossil fuel for heat, and then are concerned about polluting the air. We install scrubbing equipment to clean the smoke from the incinerator, and then have polluted water to deal with.

We see that fossil fuels are getting scarce and costly, and press for nuclear reactors to generate our electricity. But electric power accounts for only a fifth of the uses to which we put heat from fossil fuels. And engineers have yet to find a way to deal with the far-greater thermal pollution, which is lost energy, from a reactor, let alone what to do with the radio-active wastes. Merely to salvage the reac-tor fuel that does not get used on the first pass through the equipment brings an extra cost in energy, and in pollution of air, water, soil or some combination.

Each technological substitute, whether for a pollutant or a low-grade food, seems to introduce more new problems than it solves, to cost more in energy, and to require more years to a stage of actual use than can be allowed. In an advanced country, a breeding program for high-yield cereals can keep up with a modest increase in population. But it fails to be a "Green Revolution" when transferred to a develop-ing nation that lacks the necessary fertilizer, agricultural equipment, and facilities for marketing the product. To supply the fertilizer, the equipment, and the facilities (and to maintain the program) would cost more in energy and capital investment than to supply the food.

Garrett Hardin identified the central di-lemma in rational planning. His essay "The Tragedy of the Commons" (1968) compared nations, states, and ethnic groups to the members of a community who pastured their livestock on a central mead-ow, which belonged to the society rather than to a single individual. Each person gained by having his own animals eat as much as possible of the plants. Any uni-lateral move toward moderation for the benefit of all merely benefitted the few who would not cooperate. Inevitably the pasture ("the commons") was overgrazed and destroyed. Moderation is not a popular concept, as the International Whaling Com-mission found in trying to regulate the world's whalers and prevent destruction of the resource.

In placing values on alternative uses of land, civilized man puts a low rating on regenerating forest. This tree-clad area in Pennsylvania is second-growth or younger, crisscrossed by narrow roads and pocked with cellar holes where houses once stood; it produces no annual crop that can be sold, and yields oxygen and fresh water that become public property. Equally low in value, as measured in taxable production, are the pockets of primitive housing that city planners yearn to eliminate. In Hong Kong, modern structures surround a village of this kind (right) on one of the most densely populated islands in the world.

Industrial productivity in the cities, contributing to the gross national product, often seems efficient because the cost of waste disposal is not included. Steam from this plant in New York City (left) is the visible waste. Accompanying it may be invisible gases (such as sulfur dioxide from combustion of sulfur-containing petroleum fuel) that are far more damaging to the environment. Particulate matter, such as soot, is often ejected unseen when the stacks are cleaned at night. Between scows loaded with garbage from city dwellers and massive factories along the East River in New York, the occupants of a graceful sailboat try to enjoy what is left of the polluted wind and water.

173

Although pollution problems are primarily diseases of the affluent nations, solving them through moderation will hurt the rich less than the poor and the very young. Failure to solve the human population problem hits hardest at nonhuman life.

Competitive Exclusion

Each expansion of the area used by mankind and the cultigens has constricted the range of wild animals and plants. Often the survivors are limited to a few small refuge areas, having lost perhaps half a continent of former territory to the human species. The Eurasian lions disappeared, and now are represented by fewer than 500, all of them on the Gir peninsula along India's northwest coast. Bears survive in the mountains and wolves in the Far North of Eurasia and in folklore and stories for children, in whom they provoke an unreasoned fear. The Nile crocodile and the sacred ibis are gone from Egypt and the lower Nile, the grizzly bear from almost all of the coterminous United States and northern Mexico, the marsupial koala "bears" from most of the gum forests of Australia. Even on large land masses, the refuges resemble islands with confined populations, and prove equally vulnerable to change. Species that can tolerate proximity to mankind, or that benefit from human neighbors, seem mostly likely to survive.

The gravest of ecological consequences in this competitive exclusion is total extinction. Between 1 and 1650 A.D., at least 10 species of mammals and 10 of birds went prematurely to their doom, at a rate of one per 165 years. They included the ancient wild ox (aurochs) of Eurasia, the giant ground sloth (*Glyptodon*) of the West Indies, the elephant bird (*Aepyornis*) of Madagascar, and the moas (which were equally flightless) on New Zealand. No records show how many cold-blooded animals and plants became extinct in the same years from matching causes.

Between 1650 and 1850, the roster of extinct life lengthened at a rate of one more species of warm-blooded animal every five years—not every 165. The new losses included the famous dodo of Mauritius (1693) and the greak auk of North Atlantic coasts (1844). One animal that man unwittingly introduced on Mauritius helped exterminate the dodo. The newcomer was the brown rat (misleadingly called the Norway rat). Travelling aboard ships as a stowaway, it spread from Asia—possibly Shanghai—and became the wharf rat, sewer rat, and garbage-dump rat. It reached England in 1730. Larger and more aggressive than the black rat, it has been inbred as albino strains by medical men, geneticists and psychologists.

The great auk, by contrast, was killed off by fishermen who climbed to the nesting sites of the birds on offshore islands, clubbed the adults, rendering them and their young for oil, and collecting their feathers for sale to the featherbed industry. Had other sleeping equipment replaced featherbeds a century earlier, the great auk might still be a living, flightless denizen of cold coasts. These birds were the original penguins, from which the name was later transferred to unrelated inhabitants of the Southern Hemisphere.

The pace of change in the human environment and the disappearance of wildlife became fast enough by the late 1840's to impress men within single life spans. John James Audubon, who had attempted to portray in natural size and color all of the birds and mammals of North America,

When automobiles wear out, they clutter up the countryside for many more years than a dead horse—an animal that, in life, moves more slowly, eats a renewable resource as fuel, and is self-reproducing. Today horses have become almost obsolete, except for their use in racing and riding.

realized that he could never see a live great auk and that other species were dwindling rapidly in both numbers and range. After a career of shooting these animals for sport, for meat, and for models in his artistic work, he spoke out in his later years for conserving America's living resources. It seems almost anachronistic that this first realization of man's impact upon the environment should have come in the same decade that Louis Agassiz, the Swiss geologist and zoologist, discovered and popularized evidence of widespread glaciation—the Pleistocene. Today it seems that the human species as a biotic factor in the environment is having more influence on the distribution and survival of nonhuman life than the Ice Ages, which rate as a physical factor of prime magnitude. Often mankind has destroyed in less than 2,000 years what the glaciers failed to do in two million.

The rate of extinction of wild animals and plants has reached new and frightening levels. Instead of one species of bird or mammal joining the roster of vanished life each 5 years, the rate rose to one in each $9\frac{1}{2}$ months between 1850 and 1900 (64 species) and to one in each 8 months between 1900 and 1950 (76 species). In fact, of all the species that have become extinct in the last two millennia, approximately half have disappeared since the year 1900.

The losses since 1850 include, among others, the South African zebra known as the quagga (1875), the Falkland Island fox (1876), the Labrador duck (1878), the passenger pigeon (1914), the Carolina paroquet (1920), the bubal hartebeest of Algeria (1923), the Syrian wild ass (1928), the heath hen (1933), and probably the strange marsupial scavenger known as the Tasmanian wolf.

Public sentiment saved the European bison (the wisent) and the American bison ("buffalo") to become park animals; both are extinct in the wild. Similar efforts seem to have saved the trumpeter swan and whooping crane in North America, the black wildebeest and the bontebok in South Africa, and perhaps the Arabian oryx, of which the last known on native soil were shot for sport by soldiers with submachine guns.

A file of scientific information and a dossier on efforts to protect rare and endangered species has been begun under the auspices of the International Union for the Conservation of Nature and Natural Resources (IUCN) and the World Wildlife Fund, both with headquarters in Morges, Switzerland. Presently the official *Red Data Book for Mammals* in danger includes 258 species and subspecies—about 8 per cent of the world's mammals. The corresponding volume for birds totals 334. Both lists are admittedly incomplete because information on rarities is scarce and often unreliable. A *Red Book on Rare and Endangered Fish and Wildlife of the United States*, issued in 1969 by the U.S. Bureau of Sport Fisheries and Wildlife, is more complete; it lists about 15 per cent of the 395 species of mammals occurring on the United States or in surrounding waters. Some mammalogists estimate that about a third of the mammals would be considered endangered if data were more complete. Yet the United States has the most active program for wildlife management and protection on earth.

The information on birds (which are faring no better) and mammals is more comprehensive than that for cold-blooded vertebrates or invertebrates. These, however, and plants as well are now being given belated attention.

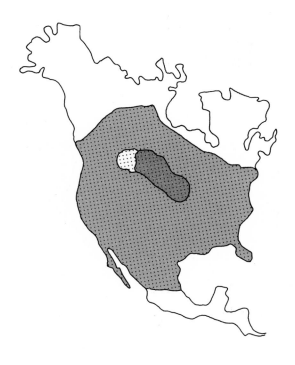

For the most clearly threatened species of vertebrate animals and seed plants, special sanctuaries are being established either on the home territory or elsewhere. Zoos and botanic gardens are finding new roles, as places in which breeding stocks of rarities can be built up, toward the day when restoration of safe habitat will permit reestablishment of these species under wild conditions. Meanwhile, in confinement, much is being learned about the requirements of these animals and plants. The new information should help greatly in restoring populations in the wild at an early date.

Ways to cherish the human environment are being tried at a rate that is often hard to appreciate because we are all so close to the ecological frontier. The program calls for sharing the land, the waters, the air, the living resources through multiple use. Sometimes it is described as providing the greatest good for the greatest number, with the indefinite goal of "perpetual" availability.

After a lapse of 6,000 to 12,000 years, attempts are being made to increase the total variety in cultigens by finding additional kinds and taking advantage of their special fitness in difficult environments. The musk ox of the Far North, which seemed headed for extinction until recently, proves to be a potential resource for meat, fine wool (called qiviut) and hides, yet is tractable as domestic cattle and magnificently adapted for life near polar coasts. In Rhodesia and adjacent parts of Africa, the large native antelope known as the eland is accepting domestication well. It makes far better use of local vegetation and the limited fresh water than the wetland cattle that man has tried to raise there for centuries. The animal is virtually immune to the endemic trypanosomes carried by tsetse flies, which cause fatal nagana disease in man's familiar cultigens. Eland meat and hides have excellent quality.

In many parts of the world, government officers are adjusting national programs to benefit from the fresh realization that vacationers with money to spend can do more for the national economy than the sale of produce from land of poor or intermediate quality. New Zealanders have turned to commercial haversting for meat and hides of European deer, whose ancestors were introduced as targets for sportsmen and became pests when their popularity waned while their numbers rose. On the remote islands of Tristan da Cunha in the South Atlantic, citizens are now "farming" the endemic birds, so as to have a surplus to share with zoos and museums in distant lands and to ensure that these once-endangered species will not become extinct.

Travel to remote areas is now easier and less expensive. Visitors from within a country and beyond gain a new perspective through personal experiences. Quickly they discover the differences in the human environment, the cultural diversities, and the intimate relationships between mankind and the natural world. These comprehensions are almost impossible to develop within the confines of a crowded city. They rarely come with adequate force to people who never travel more than a few miles from the place where they and their ancestors were born.

Mankind and social problems in cramped quarters are but parts of a living world that faces the future with us. We have companions, nonhuman but alive, meeting the same challenges among the same resources. They, like we, have a long history of past success in adjusting to change. But unlike nonhuman life, we can comprehend the path our ancestors took. Approving most of the cultural heritage they evolved, we are ready to apply it in ways that—like mankind—have no precedent. Voluntarily we can act, choosing each alternative according to its probable contribution to a long future rather than an immediate return. So far as we can learn, no other animals or plants ever had such an opportunity.

11 Interactions within the environment

"Planning with nature" has become a new goal in the use of the landscape and coastlines. The planners look more closely than ever before at the topographic maps that civil engineers have drawn up for each region. Geological features below ground, such as rock formations and subterranean drainage patterns, take on fresh significance. The nature and depth of the soil is examined. The climatic features of rainfall and temperature changes need careful consideration. The exposure of each small area to sun and wind holds special importance. Every pond and lake, stream and river is seen as a resource. Each forest and field, swamp and marsh becomes an indicator of potential uses that could be combined into a regional program benefitting both man and natural life.

Realizing that the availability of fresh water is often the paramount limiting factor in a terrestrial environment, landscape architects now tend to study one river basin at a time. They treat it as an ecosystem, isolated on each side by rocky crests. Over these crests moisture and dust, but not dissolved nutrients, may be brought by the prevailing winds. Wild animals are as likely to leave as to enter the river basin, and plant life tends to be restricted in its ability to disperse.

Within the river basin, however, human activities and geological forces are not alone in changing the scene. A field that man neglects soon disappears under a canopy of shrubs and trees. The debris from life, along with particles of mineral matter from eroding hills, fills the swamps and marshes, and eventually the ponds and lakes. These changes form a predictable succession, demonstrating the profound effects of living things on the environment.

The first scientist to describe carefully such alterations in the landscape was the botanist Anton Kerner von Marilaun, whose observations were chiefly on the plants in open water, swamps, marshes, forest edges, and adjacent grasslands. His book *The Plant Life of the Danube Basin* was largely ignored when it appeared in 1863. It found appreciative audiences and became one of the classics of ecology only after the general relationships linking vegetation to land areas were pointed out by the Dane J. E. B. Warming in 1895, Oscar Drude in 1890, and A. F. W. Schimper in 1899. That Kerner's observations had worldwide relevance was recognized by the American botanist W. S. Cooper, who discovered similar sequences of change in the mosses on Isle Royale, Michigan—an island in Lake Superior.

The orderly sequence of changes, known now as ecological succession, is easier to notice among the plants of a region than among the animals. All through the growing season, the plants can be recognized, counted, and their communities analyzed. The animals are far more elusive and all too briefly exposed to view. They tend, moreover, to associate with the plants they eat or that give them cover. This, like the trophic relationships, makes the vegetation more fundamental.

The interactions between living things and their environment take place most slowly where bare rock or water with little dissolved material awaits the colonists. This is a "primary succession" because it has no organic residues from previous residents to make the habitat more hospitable. The bare rock may be recently cooled lava or a cliff face exposed by a landslide. Water that offers so few amenities may be melted snow close to the edge of a glacier high on a mountain slope.

In all cases, the colonization of an area begins with immigration and depends upon the establishment of pioneers. Species with the most efficient disseminules from adjacent regions are the most likely to appear. For them to grow and multiply, the physical and chemical features of the environment must match the tolerances of the disseminules: temperature, moisture, light, and soluble nutrients. The first to survive on bare rock are likely to be the spores of fungi and the dormant cells of algae. A gentle rain may wash them into crevices and stimulate their germination. Some of the fungi and algae team up to form lichens that cling to the rock and grow whenever the humidity is high. The earliest of them coat the rock with a layer that is barely thicker than paint. Each lichen holds moisture between itself and its support. The moisture becomes acidic when organic materials such as fallen pollen dust decompose in it. Slowly the solution attacks the rock and creates minute pits in it. A few particles break loose and form the beginnings of soil. A surprising assortment of decomposers and microscopic animals find this microhabitat a suitable place to live. They help alter the thin soil into a medium in which different, thicker, and taller lichens can grow. The spores of mosses and ferns arrive. Little by little the community on the rock increases in area, depth, variety, and rate of soil production. Within a year the habitat may be hospitable to a seed or two if the climate is favorable or require more than a century if the weather is cold and dry most of the time.

Sometimes the force that starts a succession also brings the first colonists, as a river does in flood. More often the colonists of the land come with the wind or currents from the sea. No matter how desolate a bit of land is, it offers a resting place to birds on migration. Some drop living seeds in indigestible coats, each covered by a film of fertilizer. Strong winds of hurricane proportions occasionally transport live snails and other animals (including fishes) unharmed from place to place. Even the most unlikely colonist may arrive once in 10,000 years.

Having arrived, a potential colonist must establish itself. This too depends on chance. Many seeds lose their ability to germinate by the time they reach a suitable place or, having arrived, are wetted by a rain. Under most conditions, a majority of the seeds that germinate die before reaching maturity. A few that attain maturity fail to reproduce. This common pattern leaves to a small minority the reproduction that is essential to start a new population.

In his book *Plant Succession* (1916), the American botanist F. E. Clements expressed his own astonishment over the poor success in the establishment of seedlings from sexual reproduction as compared to the relative ease with which pioneer plants multiply asexually by means of propagules, such as runners or extensions of branching root systems. The few scattered individuals which get a roothold and then spread out asexually as clusters earned the name "family groups." Eventually, at a later stage in the succession, these family groups might expand enough to obliterate the intervening space and then come into competition with one another.

As the colonists grow more numerous, the lower green plants are shaded by the higher: the lichen by the moss, the moss by the fern, the fern by the shrub, and the shrub by the tree. The tallest gets first use of the light. Often it starves its neighbors, blocking them from solar energy until they

Life Takes Hold

Grasses and other plants that divert no energy toward development of sturdy, woody trunks are the prime colonists of drifting sand, whether in dunes along the west coast of Denmark (left) or near the north shore of Cape Cod (above). The fibrous roots form a dense matting while the upright foliage keeps the wind from reaching the thin soil. Gradually the sand becomes stabilized, and organic matter in its upper layers forms a suitable seed bed for woody plants.

181

weaken, become susceptible to diseases and die. Their remains contribute to the soil. A surprising number of kinds of plants alter their environment so much that additional seedlings of their own species can no longer get a start. Instead, the conditions fit the needs of different species, whereupon the sequence of plant succession begins. It is the inheritance of each species that determines the tolerances of the seedlings and hence, to a large degree, whether a tree that dies of old age can be replaced by another of the same species or by something else.

The Filling of a Lake

The most impressive sequences in ecological succession are those whereby a body of water is converted into land. Every lake is eventually filled in, or drained abruptly by a change in the river system of which it is a deep expansion.

The early steps are extremely inconspicuous because the colonists consist of an accumulation of microscopic green plants drifting with the water. As they die, their organic remains settle to the bottom and build up as detritus. In this layer, bacteria become abundant as decomposers. The detritus is added to at a faster rate when duckweeds, water ferns, or water hyacinths floating on the surface die and sink to the bottom. Mineral particles settle from suspension. Those that are washed from the land around the lake after a rain muddy the water for a while. The coarse particles sink to the bottom near shore while finer ones settle more slowly and tend to be carried by currents to deeper water. The river current sorts out the sediments similarly, the gravel settling first, then the sand where the current slackens, and finally the silt which builds up as mud.

Eventually the near-shore waters of a lake become wetlands which, by definition, are less than 10 feet deep and yet covered by more than 2 inches of water in all seasons. When the depth diminishes to about 6 feet, conditions are better for flowering plants to take root. Waterlilies produce loglike horizontal stems in the bottom sediments, and extend their buoyant leaves on long flexible petioles until the upper surface of each blade is exposed to air. Water celery (*Vallisneria*) grows in clumps, each leaf a long narrow pale green ribbon that stays submerged. Its inconspicuous pistillate flowers, each tethered by a spiral slender stalk the thickness of a shoelace, float upward to the surface. The staminate flowers snap off, rise, and break through the water film. By projecting into air, they catch the wind which often blows them to a pistillate flower. "Coontails" are the feathery branches of submerged hornwort (*Ceratophyllum*) or a water shield (*Cabomba*). In shallower places, the carnivorous bladderwort (*Utricularia*) grows

abundantly, and waterweed (*Elodea*) is likely to form tangles of bright green leaves among which many animals find food and refuge.

Each autumn, the great bulk of the hornwort and water shield dies and partially decomposes, while the tips of the branches sink to the bottom and lie there all winter. Like dormant buds, they are ready to begin new growth in spring. Along with the plant debris, shells of dead snails, cast exoskeletons of insects, cases of caddisworms, and feces of many kinds accumulate. Sometimes the process is accelerated in late summer because plant growth chokes the water, blocking sunlight from lower levels, preventing photosynthesis from renewing the oxygen supply there, and hastening the death of many organisms that could have survived if allowed more space. The organic matter mixes with the silt and hastens the process.

Wild rice (*Zizania*) and burreeds (*Sparganum*) grow in slightly deeper water than cattails (*Typha*) or arrowhead (*Sagittaria*)

The herbivorous beavers of north temperate lands show a special interest in poplar trees, which provide edible, nutritious bark and wood that is easily cut into manageable lengths. Some of the wood is used by the animal in building a watertight dam, holding back the flow of a stream to form a pond. The pond provides a moat around a central lodge, which the animal builds of sticks and mud as a shelter and a nursery. But a beaver pond often floods adjacent land, killing some trees and fostering the growth of others. The dam slows the stream until its suspended matter settles and the silt builds up around marsh vegetation to form a meadow. When the pond becomes too small and shallow, the beaver family moves to a new location, leaving an almost flat terrace of rich soil on the slope for the forest to colonize. When the New England colonists trapped the beavers and felled the trees, the terraces eroded quickly, destroying the basis for productive agriculture.

or pickerelweed (*Pontederia*). Each has its characteristic depth and type of leaf, but all extend their foliage and flower clusters into air. In autumn they tend to die back to their roots or buried stems, contributing a remarkable bulk of dead tissues to the materials upon which the decomposers work. Runoff water from melting snow and heavy rain drops silt among these plants and fills the spaces between them, converting the bottom to a muck that is exposed in late summer if drought develops.

Alders (*Alnus*) and willows (*Salix*) extend their roots into the muck among the emergent kinds of vegetation. Alder growth is particularly rapid once it gets a start, because actinomycete fungi in the nodules on the roots of the shrub produce nitrogenous compounds—more than the alder can use. An excess diffuses into the muck, improving nutritional conditions for vegetation of other kinds. In less than a decade, the soggy dead leaves of alder and willow complete the filling of low places that can hold water. The lake area has become a wet woodland, offering opportunities to trees of drier land. Gradually they spread in, rise tall, and shade out the alders and willows. The woodland may be flooded for a week or two each spring. But all summer its humid shade matches the tolerances of ferns that grow in dim light and wildflowers that bloom before the deciduous trees leaf out.

The Covering of a Bare Field

The succession with which most people are familiar starts with the arrival of a few weeds on a bare piece of land. We give the name weed to any plant whose disseminules disperse widely and colonize soil that man has disturbed. Many, such as the dandelion, are of Eurasian origin and have accompanied human colonists wherever they have gone, altering the landscape for human benefit, whether for agriculture, or pathways, or communities. But usually, the soil in which a weed takes root still contains organic matter from previous vegetation. The ecological succession that is beginning is a secondary one, not a primary.

Neither dandelions nor a thick sod of pioneering grasses did more than slow the growth of the native easter white pine (*Pinus strobus*) in New England during the 1830's and 1840's, when farmers abandoned the land they had cleared and moved west. Pine seeds blew in from scattered woodlots and colonized the farms. The little trees germinated readily between the granite boulders and the stones that had been left by the glaciers of the Ice Age. Field birch and staghorn sumac grew too, but soon were shaded out as the white pines rose from the fields in an almost solid stand of approximately equal age and size.

Stronger pines shaded the weaker ones and took over the space when the weaklings died. The tops formed a closed canopy, shutting off the light needed for survival of low branches; on these the needles dropped and the branches themselves broke off, leaving each pine trunk straight and knotless. By the 1880's lumbermen found a wonderful wealth in pine timber. Land owners began to dream of letting the fields produce repeated crops of these fine trees, with no need for work on them until they could be harvested in 40 years or so.

But when the tall white pines were cut and hauled away, no grass grew on the ground as a seed bed for another generation

The Succession of Life

The transformations of landscape due to successional changes in the conspicuous vegetation proceed more rapidly than is often recognized. Canals in Florida, excavated for drainage or movement of small boats, quickly become clogged with water hyacinths (below) as the first step toward shallowing and eventual filling in with soil. Around ponds, the shrubby alders spread into the shallow water (right), holding organic matter such as fallen leaves and silt from spring floods, progressively extending the land—and obliterating the pond. Fire-resistant pitch pines (far right) spread into a meadow, transforming the soil with fallen needles and resinous branches until fires become infrequent and other woody plants can expand their range in the same direction. In a decade or two, without human aid or interference, the canals vanish, the small ponds disappear, and if the rainfall will permit, forests replace the grasslands.

of the same kind of trees. Instead, the slash left by the lumbermen lay atop a thick duff of fallen pine needles. It was a fine place for squirrels to hide acorns, beechnuts, and American chestnuts. These seeds germinated, just as others had in past years. But now they had light to use as they grew through the duff. Cutting the pines gave the hardwood seedlings a chance. They, and not white pines, replaced the evergreens. The few pines that did find a place among the young oaks, beech, and chestnuts were chiefly pitch pine (*P. rigida*) and of little value.

The ecological succession in the New England community continues. The deciduous leaves from the hardwood trees mostly decompose in a year or two. They build up atop the old pine duff, becoming a rich compost in which seedlings of the same kinds of hardwoods take root easily. Whenever an old tree, such as an oak, dies and leaves an opening in the canopy, another oak or a chestnut or a beech soon competes for the light. One or two young trees fill the gap and restore the continuity of the forest.

The hardwoods differ from the conifers in another important way: after being felled, they often regenerate quickly from the stump or surviving parts below ground. Some of the resources stored by an old

hardwood tree are still available to speed the growth of new sprouts, helping them take its place in the forest. So vigorous was this schedule of replacement that after 1909, when an introduced fungus disease called chestnut blight essentially wiped out the American chestnut east of the Appalachian Mountains, other hardwoods filled the gaps left by this once-abundant tree. Because a hardwood forest with its small admixture of conifers, such as hemlock, can replace itself indefinitely if left alone, it is regarded as the climax community in this particular ecological succession.

One Climax or Many?

For each part of the terrestrial world, the climatic pattern of temperature and precipitation determines to a large extent what kinds of plants will be able to form a climax community. Commonly the most conspicuous vegetation is referred to as dominant, and gives the community a useful name (such as an oak-beech-maple-hickory community). On level land the dominant plants may show remarkable uniformity over many square miles. In hilly or mountainous country, the dominant community is modified in understandable ways according to the detailed topography. In the North Temperate Zone, for example, a north-facing slope receives less sunlight

than a south-facing one; a west-facing slope is likely to get more rain than an east-facing one. Generally the climate on the southwest slope is sufficiently milder than on the northeast slope to produce a measurably different climax community. The milder one will have fewer conifers and resemble more a community on level land nearer the Equator, while the other will have more conifers and correspond to plant communities farther north.

If allowance is made for these climatic variations within a region, and the dominant plants are viewed alone, it is often possible to conclude that a single climax community develops in the region, regardless of whether the successive stages began with the filling of a lake or the covering of bare earth. Ecologists call this a monoclimax. But when the lesser plants and animals are considered, it becomes evident that differences in the abundance of individual species bear less relation to the direction of slope (and hence microclimate) than to whether the succession developed on bare sand rather than on rock, clay soil, or in a pond. Some of these differences can be recognized among the dominant trees too, even in an aerial photograph, because they match variations in soil chemistry and the drainage pattern. Where these differences in soil and land structure affect the living community, the differences are perpetuated in many climaxes—that is, a polyclimax.

During a century or more, the natural ecological succession is often stalled or set back repeatedly to earlier stages (called seral stages) by the introduction of new species or by human activities. In New Zealand, so many exotic animals have been set free and are still extending their ranges that ecologists have yet to find an ecological community they can regard as having reached the climax stage. Even the tussock grasslands on the mountain slopes and the dense forests of antarctic beech (*Nothofagus*) seem to be proceeding through seral stages in unpredictable ways that have no apparent end. Scientists look in vain for the kind of stability that is recognized in the Northern Hemisphere, where particular types of forest or prairie grasses grow as the climax vegetation.

In some place, cyclic successions with no real climax are now known. In the Scottish highlands, the shrubby heaths and heathers grow close together, shading out other vegetation. Eventually, however, the shrubs grow old and die, exposing bare stems. On these a lichen (*Cladonia silvatica*) forms a dense covering and becomes the temporary dominant. Then the stems crumble, letting the lichen disintegrate, leaving bare soil. This is soon colonized by bearberry (*Arctostaphylos*). The pioneering bearberries are gradually invaded by heaths and heathers, which re-establish their cyclic dominance.

Similar cyclic changes occur among the oyster beds in shallow coastal waters. The adult oysters appear able to inhibit the larval (spat) stages from settling at a rate faster than is needed to maintain a continuous coating over the bottom. But blue mussels (*Mytilus*) gradually invade and smother the oysters. In turn, acorn barnacles (*Balanus*) settle on the mussels and form a crust so thick that the mussels starve. But soon after the mussels die, their thread-sized tethering strands disintegrate. Storm waves tear away the mussel shells and barnacles together. Only a hard coating of dead oyster shells is left on the bottom. Young oyster larvae settle on them and start the ecological cycle once again.

Setting Back the Succession

Both natural factors and human activities can destroy a natural community, setting back the ecological succession to an early stage. Each year, volcanic action and major fires affect many areas, some measuring thousands of acres. Landslides tend to be more local. Glaciers advance slowly as well as retreat. Those of the Ice Age that began less than two million years ago erased all life from more than half of North America and much of Europe. They set back the previous successions in the greatest destruction of habitat known since life colonized the land. From this event, Greenland and some other areas in the Far North and at high altitude have still not been able to recover.

Volcanic activity may sear an area with a flow of molten lava, or bury it under a layer of particles ranging in size from ash to cinder, lapilli, bombs, and blocks. One vast lava field of Pleistocene age in the Columbia River region of the western United States measures more than 100 miles by 50 miles, and has an average thickness of lava approximating 400 feet. The volume of extruded basaltic matter is close to 400 cubic miles. This is more than 100 times larger than the largest lava flow that man has observed in action. The Columbia lava field is still far from being covered by vegetation.

Biologists have had an opportunity to follow the successional changes in the recolonization of volcanic areas in several parts of the world. The first to be studied in detail was the island of Krakatau, near Java, where the volcanic eruption of August 26–27, 1883 destroyed all life. Lava flows in the eastern Congo, and in Hawaii more recently, have offered more convenient examples for scientific investigation.

Sometimes the volcanic ash that settles on a mountain slope can be dangerous. Heavy rain can convert it into a fluid mud that suddenly flows down, sweeping away forests and causing immense damage before coming to rest near the base of the volcano. Yet when ash is worked into the soil, as has happened naturally in many parts of Central America from Guatemala to Costa Rica (excepting Honduras), it provides a coarse and beneficial texture. It helps retain water that otherwise would drain downward or run over the surface, carrying away soluble mineral nutrients. Ecologists in Honduras complain that their soils are poor because they have had no volcanic activity in recent times, while agriculturalists near volcanoes in Costa Rica, Nicaragua, El Salvador, and Guatemala bewail the crops that are lost where volcanic ash is falling from new eruptions.

Fire, which lightning may kindle, can level a forest and burn the organic matter in the upper soil, leaving an almost sterile surface of mineral particles. Or the fire can remain in the crowns of the trees, killing them, but doing little more damage at ground level than to remove some of the dry litter of fallen foliage and branches. Both types of destruction let light reach the soil, and temporarily enrich it by soluble materials from the ashes, especially calcium, potassium, and compounds containing phosphorus and nitrogen.

If the roots below ground have been destroyed and nonliving organic matter too, the water-holding capacity of the soil is seriously impaired. Erosion frequently follows until a new covering of foliage and new extension of roots gives protection. Many of the nutrients may be lost quickly through leaching. These conditions, however, suit quite a few herbaceous pioneers, such as fireweed (*Epilobium*). These plants shield the ground in a year or two while the surviving shrubs and trees regenerate by sprouts, and new seedlings compete for a place in the sun.

Fire has prepared large areas of North America as a seed bed for forest trees. Without fire, in fact, the great stands of Douglas fir, western white pine, and other trees in the Northwest cannot perpetuate themselves. Nor can those of longleaf and loblolly pine in the Southeast, or western yellow pine in the western mountains, or red (Norway) pine and jack pine in much of Canada and the United States. Their cones fail to open and release the seeds inside until after the seeds have died of old age, unless a hot fire scorches the cone scales and causes them to curl apart. Foresters are beginning to realize that complete control of fires actually promotes the succession of trees of lesser value to a climax forest. Fire has become an essential factor in the maintenance of certain habitats, and to animals that live there. An example is the Kirtland warbler which nests only among young jack pines in northern Michigan; it is now so rare that ornithologists call it "the bird worth a forest fire."

In parts of Alaska, fires during the past century have destroyed the white spruce and black spruce forests over enormous areas, and also the shrubby lichens that grow slowly on the thin soil. Some of these communities have changed to areas dominated by fire-tolerant grasses and sedges, or by low-growing willows and dwarf arctic birches. The time required for the original climax to restore itself is almost unbelievably long due to the short growing season each year and the tremendous distances that disseminules of climax dominants have to travel.

All over the world during the last 30,000 years, and especially during the most recent 8,000, climax communities have been invaded by mankind. Accidentally or deliberately, people have started fires, setting back the succession. Almost always the change is toward conditions with less water and harsher weather at ground level than characterized the previous climaxes. In many places, the area of man-altered landscape is so broad that, when farm land is abandoned, the former natural communities re-establish themselves very slowly. The native plants are neither near nor numerous. Colonists have far to travel before they can compete with the weeds. The native flora and its associate animals reach first the rim of the abandoned land, making it the zone that shows the earliest changes toward the maturity of climax. This pattern is typical in all successions, with the central region the last to reach the climax stage.

With modern equipment, man can change the ecological character of large areas in just a few years. He impounds water in new lakes behind big dams, perhaps converting an area of forest into a place for planktonic life. Storage of the water behind the dam increases its transparency to light, and affects its temperature. It tends to be warmer in winter and cooler in summer as it flows on down the river from beneath the controlling dam. This may improve the artificial lake as a home for water plants and fish. The alteration of the scene is a massive counterpart of the ponds produced in the Northern Hemisphere for millennia by beavers with their log-and-mud dams. Yet every dam, whether beaver-built or man-made, eventually collapses. The impounded water drains away. Silt that settled for years behind the dam becomes

Volcanic activity within the last few centuries has left craters, cones, black lava, and cinder in a tract of almost 100 square miles in south central Idaho, now the Craters of the Moon National Monument. Only a very few species of plants have been able to colonize the devastated area so far. The trees on the hillside in the distance are on a natural ridge that escaped the full effect of the eruption.

fertile land where previous erosion gouged out the substratum to form a valley. Succession resumes.

When man fills a wetland or levels a hill, the change is rarely a prelude toward letting climax vegetation occupy the area. More often the leveled land is turned into a housing development, an industrial park, an airport, or a section of roadway for high-speed travel. It may be paved, covered by buildings with impervious roofs, or converted into a grassy tract that is kept mowed and as free as possible of invading weeds. But human plans change, particularly if they originally were made without regard for nature. Eventually these lands become available to living things that tolerate the sun and wind. As the products of man's construction fall into disuse and disrepair, the pioneer plants and the animals that consume them begin a new succession of some kind.

It is possible, of course, for man to undertake an afforestation program, setting out trees where none have grown for years. Generally undertakings of this kind are planned toward a crop of lumber. The young trees of a single species are planted in neat rows, to be harvested long before they reach old age. Rarely is the intent to clothe a mountain slope or river basin with a suitable mixture of species that could soon reach the steady state of interaction characteristic of a climax community, even though this would improve the water-catching and -holding power of the entire area.

The ecosystem that can perpetuate itself indefinitely has probably reached the most efficient pattern of utilizing nonliving resources under normal climatic conditions. It has minimal effect on adjacent ecosystems. But it has nothing that man can harvest without destroying it, because all of the energy captured in a year goes to the combined respiration of the plants and the heterotrophs. No husbandryman or domestic animals can use the climax community more than once. Their needs are met best, instead, when few species are present and the simplified ecosystem is easy to control. A crop of grains, timber, or livestock is taken as soon as its growth is nearing completion. The product of the land is hauled away, while the organisms that maintain and produce the soil are ignored. In man's planning for maximum yield year after year, there is no place for an old herb, an old tree, an old animal, or an old community. When the crop is taken (or destroyed by fire or disease), the land is essentially bare—ready again for the pioneers. Only under these simplest of conditions is the yield high.

The Boundary between Communities

The most unstable environments, where natural colonization succeeds for a while and then fails for a time, lie along the boundary between communities. The pioneers from both realms compete there and enrich the variety of life by their presence. The shifting boundary between open water and a marsh or a swamp, or between a forested area and an adjacent grassland, is especially attractive. Living things in these situations can often benefit alternately from two different facets of the world.

By going among the marsh plants, waterfowl can hide from hawks and find seclusion for their nests. Then, by traveling just a few yards, these birds reach a food supply of aquatic plants and animals. Similarly, deer and rabbits take shelter in the shady forest by day, and venture each night along the edge or into the grasslands where edible foliage grows within reach.

The array of life along the boundaries of major communities affords ideal hunting for predators. Many of the beasts that haunt these edges have handsome pelts, such as mink and otter along the wet edges, foxes and weasels (ermine) along the dry edges. Trappers and hunters knew about the ''edge effect'' long before ecologists gave it a name.

Many an edge goes unnoticed because it forms an irregular ring of small size around some rock outcropping or local area of peculiar soil or drainage. These isolated pockets within a climax community are havens for species of plants and animals that have utterly vanished from adjacent land in all directions. Despite competition, species that become rare under natural circumstances can hold on for centuries or millennia without disappearing completely. Their populations survive, reproducing themselves, always ready for any change

that may set the succession back a step and once more provide the conditions to which they are especially well adapted.

The adaptations we notice most among plants are those that help them survive as hydrophytes (''water plants'') with many of their parts immersed in water, or as mesophytes where the amount of water is moderate, or as xerophytes where drought is chronic. Each of these styles of plants possesses an interlocking set of features that meets the challenges in the ecological situation where the plant is found most commonly. We think of these as adaptations to life in aquatic habitats, in forests, or deserts. Yet they are related even more fundamentally to climate. On land, particularly, the climate and the climax community match one another, each with some influence on the other. Although much modified by man, they still interact in ways that lend character to each area of the world.

12 The marine world

An abundance of water in liquid form makes our world unique in our solar system. Had emerging man realized that this water covers more than 70 per cent of the planet's surface, he might better have named the world Oceanus instead of Earth. About 329 million cubic miles of water fill the ocean basins. The average depth below the sea surface is about 12,600 feet. It would still be about 8,340 feet deep if all of the solid matter of the planet were level instead of being raised above the sea surface into continents and islands.

The marine world came into existence about $4\frac{1}{2}$ billion years ago, and provided life with its first living space. The seas remain the key to life's present and future, serving as they do as a balance wheel on the climate and as the reservoirs from which the great hydrologic cycle transfers fresh water to the continents, supporting life on land. Although no one knows exactly when the earliest living things evolved, abundant evidence points to their having availed themselves of the water itself, the dissolved nutrients, the buoyancy, the thermal inertia, the protection from deadly solar radiations, the swirling currents, and the tidal changes that keep oceans everywhere from stagnating.

At least a token of every phylum of microbes, plants, and animals that ever evolved still lives in the marine world. These organisms occupy a striking array of habitats near shore and in open water, between the surface and the bottom, the Equator and the polar regions. Since the oceans are all interconnected and apparently always have been, the barriers to dispersal of marine life have been persistent currents, low temperatures, excessive pressure of the water as depth increases, and an unsuitable supply of energy.

So vast a realm without obvious internal boundaries offers few logical subdivisions. The topmost 60 feet, however, harbor most of the green plants—the producers of the marine world. Between 60 and 600 feet, the light becomes too dim to support photosynthesis, although animals of many kinds can still detect daylight and use it for orientation and for timing their activities. Six hundred feet happens to be the average depth at the edge of the continental shelves. Above this level lies less than a twentieth of the total volume of the oceans. Yet it is the only part in which solar energy is captured, serving plants with chlorophyll, and indirectly most of the marine animals and some of the decomposers.

Nearly nine-tenths of the total area of the marine world qualifies as open ocean. It is deep and remote from land. The other tenth is close enough to continents and islands that oceanic currents are deflected. This and run-off water from the land brings dissolved mineral nutrients into illuminated levels of the sea, improving the chemical environment so much that green plants are able to grow and reproduce better than in the waters far from land. Except in regions where the plants receive a subsidy in nutrients from the bottom or from terrestrial ecosystems, their growth is regulated by local decomposition and their total productivity each year is surprisingly low.

The Illuminated Depths of the Open Oceans

Surface currents, mostly propelled by prevailing winds, carry the illuminated upper waters of the oceans from place to place. Drifting along, with no ability to swim consistently in any horizontal direction, are countless minute plants and animals referred to collectively by the ecologist as

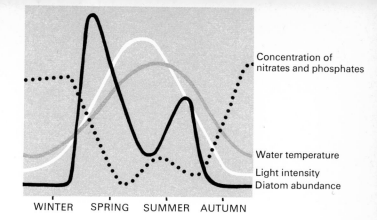

The abundance of diatoms in plankton of the Atlantic increases abruptly in spring as solar energy increases. As nitrates and phosphates are bound up in the plankton organisms, their concentration falls and causes a sharp decrease until decomposers release new supplies in late summer.

Concentration of nitrates and phosphates

Water temperature
Light intensity
Diatom abundance

WINTER   SPRING   SUMMER   AUTUMN

plankton, from the Greek word for wanderers. The plants in the plankton are phytoplankton, and the animals zooplankton. These organisms are the food base for larger forms of life, such as fishes, sea turtles, sea birds, and whales, all of which can travel with better control over horizontal direction.

If all of the phytoplankton in a vertical column of water far from land were collected on a single day and compressed to the density of cells in a green leaf, they would produce a film scarcely as thick as a sheet of paper. This seems paradoxical in an ecosystem where water is abundant, light is adequate, and carbon dioxide as a raw material for photosynthesis comes plentifully from the respiration of green plants at night, and of the animals and decomposers at all times.

The limiting factor is dissolved mineral nutrients, especially nitrates and phosphates. The productivity of the open sea for animals that man might use as food is further reduced by the microscopic size characteristic of the green plants. Most of them are diatoms, whose largest dimensions range from a thousandth to a five-thousandth of an inch (25 to 5 microns). They are too small to be eaten by most herbivores big enough to be seen easily, let alone to be eaten by man. Instead, the common herbivores of the open sea are protozoans and the larvae of minute crustaceans. The unicellular types range in size from tiny coccolithophorids and vase-shaped tintinnids to the larger radiolarians and foraminiferans less than $\frac{1}{8}$ inch across.

Until recently, the principal herbivores were assumed to be the copepod crustaceans among the zooplankton. Copepods sometimes are as big as a grain of rice. But careful study of their mouthparts showed

that the true trophic level for most of them as adults is that of first-order carnivores. The larger crustaceans, small fishes and other marine animals that eat copepods are actually second-order carnivores. Larger fishes and mammals that man harvests from the sea are still more remote from the producer level in the pyramid of marine life. Their numbers are comparatively few because they depend upon energy that has been transferred so many times.

In some regions, the initial productivity of the phytoplankton supports immense populations of shrimplike crustaceans that whalers know as "krill" (Euphausia). Before the dissolved mineral nutrients that decomposers freed during the winter are used up by the phytoplankton, the krill swim about, filtering out both the phytoplankton and the minute zooplankton. Some krill attain a length of two inches, but most are mature at an inch or less. Whalebone (baleen) whales have long followed migratory routes that brought them to northern waters in time to feast on krill, and then to the waters around Antarctica by December when the southern latitudes got their share of daylight. During the rest of the year the krill seem to feed on or near the bottom.

In other regions, the ocean currents bring up dissolved mineral nutrients either seasonally (as in the Arabian Sea during the monsoons) or throughout the year (as along the Atlantic coast of South West Africa and the Pacific coast of Peru). This subsidy supports tremendous numbers of phytoplankton and then a pyramid of consumers and decomposers. Off the Peruvian coast in the cold waters of the Humboldt current, the food chain is particularly short because the diatoms include so many that are colonial. They cling together in a transparent matrix which they secrete. Masses of the green plants from 1/16 to 1/2 inch are not unusual. The plants are eaten directly by small fishes, most notably the 4- to 6-inch anchoveta (*Cetengraulis*) upon which the cormorants known as guanay birds have so long depended. Anchovetas sustain also the pelicans, the sea lions, and other fish-eaters in the marine community.

Presently, fishermen compete with the guanay birds. This fishery, together with those in other regions of upwelling, yields almost half of the total commercial harvest from the sea; most of this half is converted into fish meal, and used as a protein supplement for domesticated livestock in the developed countries. The other half of the annual harvest comes from coastal waters nearer the major markets for fish to be used directly in human diets. The open sea offers little extra.

Vertical Migrants

Since the late 1940's, many fishermen have equipped their boats with electronic devices called echo-sounders, which send short pulses of high-frequency sound from the ship's hull to the bottom and detect the echoes from the sea floor. The interval between a pulse and its echo gives a measure of the depth of the water. The most modern devices of this kind serve also in locating schools of fish. But over deep water, the echo-sounder often indicates a "phantom bottom" that cannot be confirmed by the old technique of lowering a line with a lead weight on the end. Moreover, the "deep-scattering layer" (DSL) that reflects the sound waves rises to the surface at night and descends by noon to a depth of as much as 600 feet. Its nature is not yet completely known. However, sam-

Pounding rhythmically against a rocky coast, waves bring energy from afar. Their impact upon the land varies according to the direction of the wind, the stage of the tide, and the damping action of heavy kelps, whose gas-filled floats buoy up rubbery fronds tethered by holdfasts as much as 60 feet below.

pling of the sea at appropriate depths and observations made by explorers in midget submarines convince ecologists that the DSL is a concentration of animals. Included are two-inch lantern fishes (myctophids) with air bladders that reflect sound waves, predatory squids in remarkable variety, krill crustaceans and copepods, arrowworms, deep-sea medusae, and colonial coelenterates (siphonophores) in extraordinary abundance. Regularly each day the whole food web of animal life travels down and up again.

Even when the reality of the daily vertical migrations between the dark depths and the surface was established, the adaptive advantage in expending energy for so much locomotion remained a challenging mystery. Scientists looked for harmful effects of light on the many carnivores that follow this routine, but found none. They sought to prove that the drifting green plants (the phytoplankton) produce toxic materials while engaged in photosynthesis, and that the animals dive to escape being poisoned. But sea water dipped up from the surface by day and dense with phytoplankton in full activity has no lethal effect. In fact, it is richest then with oxygen, for the combined photosynthetic activities of so many plants produce a daily "oxygen pulse."

Recently the distinguished British oceanographer Sir Alistair Hardy offered another suggestion. He noted that surface waters are shifted in many directions, whereas deeper waters tend to flow in constant currents. He thinks of a copepod crustacean swimming vigorously downward before dawn, out of one hunting area. All day at a depth of 300 to 600 feet, it would be carried horizontally by some current. After sunset it would swim vigorously upward into a completely different patch of surface water. In one night it could not deplete the supply of herbivorous animals in one patch, or starve to death if the next night it found almost nothing to prey on. Every night it would gain a new opportunity, and a far better chance for survival, growth, and reproduction than could be found by swimming horizontally 600 to 12,000 feet in each 24 hours, all in the same area of water near the surface.

The Dark Abysses

Below 600 feet, the ocean is dark and affords nourishment to only a sparse population of swimming animals. They scavenge for the dead bodies, fragments and feces that sink from higher strata. Much of this material settles to the sea floor, upon which scavengers, detritus feeders, and decomposers are more numerous. Beyond the continental shelves, the bottom itself slopes down more rapidly to about 6,000 feet. In about 20 places, it dips below 18,000 feet in the form of extraordinary trenches, of which the deepest is the Marianas trench in the Pacific Ocean near Guam, with a recorded maximum of 35,958 feet below sea level.

Only bacteria and fungi represent the plant kingdom in these dark abysses. The animals are highly adapted, chiefly fishes, squids, crustaceans, and medusae. The carnivores eat the scavengers and one another. This far from light, theirs is the most meager diet in the sea. Correspondingly, their numbers are truly small although the variety among these abyssal denizens is remarkable.

The special prizes are the occasional large masses of animal tissue that descend silently. A mass an inch or two in diameter could well be the meal of the month for

In shallow water, the subtropical and tropical sea urchins of genus Toxopneustes *(top, left) often carry bits of mollusk shell, coral, and seaweed above their bodies, as though shielding themselves from excessive light. In deeper water, where the solar energy has been filtered and considerably absorbed, the urchins move about unencumbered. At mating season, a female horseshoe crab* (Limulus) *attracts males, which clip themselves to the back corners of her shell and force her to drag them along until she is ready to lay her eggs in the sand. The chemical attractant from some females is so powerful that it induces additional males to clip onto the shell of the first mate (left). Hermit crabs (above) conceal their soft abdomens in empty snail shells of suitable size and spend much time inspecting potential replacements before committing themselves to change. Carrying its portable shelter, the hermit crab may venture out on land and scavenge at a considerable distance from the beach. Always it is ready at the first disturbance to draw back into the snail shell.*

most of the scavenging fishes in the depths. Adaptations that help in detecting food in the darkness and in engulfing pieces quickly have great importance. Fins are extended commonly into long tactile processes analogous to antennae, which must serve also in warning of the approach of predators. Mouths, gullets, and stomachs are extraordinarily expansible, capable of accommodating food as large as the body of the scavenger itself. These adaptations confer upon the deep-sea scavengers the fanciful appearance of miniature dragons.

Most, if not all, of the animals swimming about in the dark depths are related more closely to denizens of the illuminated waters above than to those of the sea floor or the coastal province. Many of the fishes in the abysses still release buoyant fertilized eggs that rise to the surface and that hatch there as young with little resemblance to their parents. Those that survive, eating small animals in illuminated waters, gradually transform (metamorphose) and descend to the dark abysses for which they are by then so well fitted.

The Life of the Sea Floor

The ocean bottom receives a steady rain of organic particles in many sizes. They maintain a sedimentary layer that is often many feet thick, throughout which decomposition bacteria are active. These organisms, and the animals that feed more superficially upon the organic matter are a strange lot. Most are known as barophiles because they not only tolerate the great pressure caused by the overlying water but cannot survive if raised to the surface of the ocean where only atmospheric pressure prevails. The barophilic bottom animals include annelid and echiurid worms that burrow through the sediments; sea urchins, sea stars, sea cucumbers, crustaceans, and mollusks that dredge the oozy material for nourishment; sponges, beard worms and bivalves that maintain their own water currents in and out while capturing bacteria and particles of detritus; sea lilies that wait with outstretched arms for particles to settle on the upper surfaces; sea anemones that spread their tentacles for unwary animals; and fishes that prey upon whatever edible they can find. Even before they can be brought to the surface in a dredge and transferred to a pressure chamber kept at a temperature close to 32 degrees F (0°C), these animals usually die, often turned almost inside out through their own mouths.

In the deepest trenches of the oceans, the animals show evidence of special isolation and are referred to as "hadal" forms of life. Nearly 300 species have been collected at depths greater than 20,000 feet. These creatures seem most related to bottom dwellers near shore in shallow waters. Curiously, no decapod crustacean, lamp shell (brachiopod) or turbellarian flatworm has yet been found so far down.

Everywhere the sea floor is covered by sedimentary material. It is called clay if it contains less than 30 per cent of recognizable skeletal remains of animals and plants, but ooze if the proportion is greater. Unconsolidated oozes cover almost 64 million square miles, which is about 60.6 per cent of the sea floor. Commonest is *Globigerina* ooze, named for the limy shells from foraminiferans of this genus which give bottom samples a milky white or rose or yellow or brown color. It is the characteristic sediment below most of the Atlantic oceans, some of the Caribbean Sea and Gulf of Mexico, the eastern and western South Pacific, and the western Indian Ocean. Least common, but equally limited to depths less than 10,000 feet, is pteropod ooze, in which the limy shells of pteropod mollusks confer a white to pale brown color, tinged with pink or red or yellow. Pteropod ooze occurs mostly in the middle of the North and the South Atlantic Oceans.

Diatomaceous ooze is commonest at a depth of 12,800 feet, and generally is straw-yellow or cream-colored. It forms the bottom surface in waters near Antarctica and also in the northern North Pacific Ocean. Equally siliceous but less widespread is radiolarian ooze, which varies from straw-colored to red and chocolate-brown. It is rare at depths less than 14,000 feet and most abundant around 17,400. The largest area of radiolarian ooze stretches from the longitude of Hawaii almost to the coast of Central America, just north of the Equator in the North Pacific Ocean.

The clays cover almost 40 million square miles, all at depths greater than 14,000 feet. They are brick-red in the North Atlantic, but bluish or chocolate-brown in the South Pacific and the Indian Oceans. Mostly they consist of volcanic dust remaining after calcareous and siliceous materials have redissolved, which they do in cold water at great depths. Limy materials disappear below 12,700 feet, except for shark teeth and whale ear bones (composed of insoluble tricalcium phosphate), which litter the sea floor. (See diagram on page 205.)

Sedimentary particles washed from the land cover the continental shelves and part of the continental slopes beyond. The accumulation is broadest and thickest around the mouths of great rivers, narrower and thinner in between. On and in these sediments live many kinds of animals. The

202

The snail shell that a hermit crab carries about (top) offers support to a sea anemone and shelter for a segmented polychaete worm whose body can be seen projecting above the crab's legs at the extreme right. Both the anemone and the worm are commensal companions, sharing in fragments from the crab's meals. The bivalve mollusks (bottom) benefit from their association with corals by getting particles of food from the tidal currents.

variety increases greatly in the shoreward direction.

Neritic Waters

Seafaring men have long referred to "inshore" parts of the ocean as those along a coast, and "offshore" as those beyond. Ecologists see a meaningful subdivision of the marine world where the sea floor is close enough to the surface for green plants to grow on it and where dissolved nutrients arrive along with sediments from the land. They speak of this zone around the edge of the sea as the neritic province, as contrasted with the oceanic province beyond. Often the boundary is doubly visible on the bottom, both because it is the outer edge of the seaweed forests and because the slope of the sea floor changes. Seaward, the bare surface of the continental shelf slants toward its rim at an average rate of only 12 feet to the mile, which is too little to notice. Shoreward, however, the slope becomes perceptible, maintained by the force of storm waves and coastal currents.

The neritic province is the most changeable part of the sea. Geological forces continuously modify its extent and location. Currently it includes narrow fringing reefs around many volcanic islands in the Pacific world, and an expanse 800 miles wide in the Arctic Ocean along the coast of Siberia. During the past 15,000 years, the neritic province has shifted downward on some coasts and up on others. Sea beaches of New Guinea have been elevated 600 to 800 feet above sea level. Land as much as 200 miles wide between Nova Scotia and North Carolina has sunk below the depth at which seaweeds can carry on photosynthesis. Territory on which people lived

The sea floor coverings

Type	Area in millions of square miles	Percentage of sea floor	Average depth in feet
Deposits of land origin	26.6	25.7	3,000
Calcareous oozes			
Pteropod	0.6	0.4	6,800
Globigerina	48.8	35.1	11,800
Siliceous oozes			
Diatomaceous	11.9	8.6	12,800
Radiolarian	2.7	1.9	17,400
Clay	39.4	28.3	17,800

is now 200 feet under the waves of the North Sea, the Irish Sea, and the English Channel. Some of these changes can be attributed to the melting of Ice Age glaciers, and others to adjustments of continents that had been warped by the weight of the thick ice. Still others seem due to vertical movements of the earth's crust for which no satisfactory explanation has yet been found.

Even without these alterations in the relative levels of land and sea, the water may rise and fall through regular tidal movements produced by the gravitational pull of the moon and sun. In the Mediterranean Sea, this variation is so slight that it was overlooked by even the most observant men of classical civilizations. In the Bay of Fundy and along some parts of the French coast, the range between high tide and low is more than 40 feet. At the Pacific end of the Panama Canal, the tide rises and falls more than 12 feet, whereas at the Atlantic (Caribbean) end just 40 miles away, the tidal change is less than 3 feet.

From the limit of high tide to a depth of 120 to 180 feet, depending on the trans-

parency of the water, the neritic province possesses a wealth of coarse plants and clinging animals in a distinctive sequence of ecological communities. Their distribution corresponds to the nature of the bottom and its slope. The steepest slopes are those where rocky shores rise out of fairly deep water. The gentlest are those where the coastline provides protection from the full force of storms, and soft muds consolidate slowly. Intermediate are stony and sandy slopes, often with rocky cliffs or sand dunes immediately inland. Each shore form requires special adaptations in the organisms that live there.

In the neritic province of tropical waters, and as far from the Equator as the temperature of the surface water averages 70°F, rocky bottoms support outstanding communities of coralline plants and animals. They build and maintain great reefs of porous limestone. Those of the Atlantic Ocean, the Indian Ocean and adjacent seas (such as the Caribbean, the Mediterranean, and the Red) are produced to a large extent by coral animals, whereas those of the Pacific are generally dominated by coralline

algae. Charles Darwin offered the first plausible hypothesis to account for the fringing reefs around volcanic islands and the atolls surrounding a shallow lagoon. On later expeditions, scientists were able to confirm that the reef organisms protect the volcanic rock from rapid erosion under water, and perpetuate themselves as a ring of low islands after the volcanic peak crumbles or subsides. Between the islands, a central lagoon remains with a floor of live organisms.

Recently biologists discovered that reef-forming corals need the assistance of their associated algae, and deposit lime only while the algae are engaged in photosynthesis. So interlocked are the lime-secreting activities that growth lines in the coral rock record daily differences as well as seasonal changes. Fossilized corals extend the history backward, and prove what had been suspected from other phenomena: the number of days in a year is decreasing, due to a slowing of the earth's rotation by tidal friction. In Cambrian times, the year had 425 days, during the Carboniferous 390, in early Eocene 371, and since about 10 million years ago 367 or less.

At any latitude along a rocky shoreline, the transition between marine and terrestrial environments is often so abrupt that the seaweeds and sea stars live quite literally within a "stone's throw" of a forest or a grassland. The rock wall deflects most of the salt water during storms, protecting the land vegetation from intolerable amounts of salt. Often the marine vegetation is almost as effective in damping the vigor of wave action and protecting the shore from erosion. This is most noticeable where the coarse kelps form coastal forests under water near shore around the southern

The torpedo-shaped cuttlefish (Sepia officinalis) *(top, left)* of the Mediterranean uses jet propulsion to dart backward through a school of fish while it uses its trailing arms to seize a victim. If pursued, the animal can change colors spectacularly and also eject a cloud of ink. The bottlenose dolphin (Tursiops truncatus), *which is a small toothed whale, uses echolocation with ultrasonic chirps to locate and follow the fishes and squids upon which it feeds (left). As it swims among the coral reefs, the moorish idol fish* (Zanclus cornutus) *of the central and eastern Pacific is almost invisible because of its disruptive color pattern (above).*

Waving winglike extensions of the body on each side, manta rays *(overleaf) propel themselves as though flying through the water. Since they feed on plankton, they offer almost no competition to the many bony fishes that school in the same regions.*

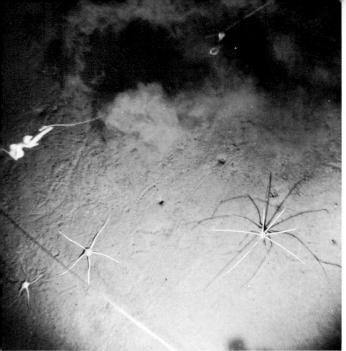

The sea floor may be far below the surface and dark, except for the light produced by living organisms. It is freshened by currents that leave ripple marks, such as those along the south side of the Java Trench at a depth of 10,000 feet (left). Worms burrow in the bottom sediments, among which decomposers of many kinds are active, nourished by an almost steady rain of organic particles from above. Long-legged sea spiders stride over its surface (top), while fishes such as flounders lie flat and inconspicuous, waiting for edible animals to come swimming within reach (above).

tip of South America and of Africa, along the cool edge of Australia and New Zealand, and northward from California and from New England to regions with similar water temperatures on Asiatic and European coasts. Some kelps grow in 60 to 80 feet of water, with immense holdfasts embedded in the sea floor, and long thick cylindrical stalks stretched to the surface; there their flat branching blades undulate with the waves and trap solar energy.

On large rocks that are submerged at high tide, different seaweeds maintain a firm grip with rootlike holdfasts while their leathery blades yield to the moving water or lie limply, unsupported, exposed to air while the tide is out. Some, such as the rockweeds (*Fucus* and *Ascophyllum*) in cool waters of the North Temperate and the Arctic, and Neptune's necklace (*Hormosira*) in corresponding waters of the South Temperate and along the shores of subantarctic islands, have gas-filled bladders that hold their fronds well separated and toward the light when the tide covers them.

Healthy kelps and rockweeds seem to keep their blades and holdfasts clean by producing antibiotic substances that inhibit the growth of epiphytic coverings such as diatoms and bacteria. A few kinds of microbes do maintain an attachment, but extend away from the alga's surface almost at right angles, where diffusion into the surrounding sea water decreases the concentration of the toxic material. A few snails feed on the algae. The principal herbivores appear to be sea urchins. In a few places along the California coast in recent years, the kelps have proliferated amazingly, matching the decrease in populations of sea urchins caused by a rise in the number of sea otters (which feed mostly on sea urchins and abalones).

A sampling of small creatures known collectively as marine plankton (bottom) may include a fish embryo still inside its transparent envelope (far right), a slender arrowworm 1 to 2 inches long (on the diagonal), a crab larva with big compound eyes and an enormous spine from its back (below the arrowworm), a small jellyfish (above the arrowworm), a number of rice-sized copepod crustaceans, and some minute green plant cells, diatoms. Only a specialist is likely to recognize the embryos in the top row: the veliger larva of the mud snail Nassarius (right); the auricularia larva of a burrowing sea cucumber (center), and the echinopluteus larva of a sea urchin (far right).

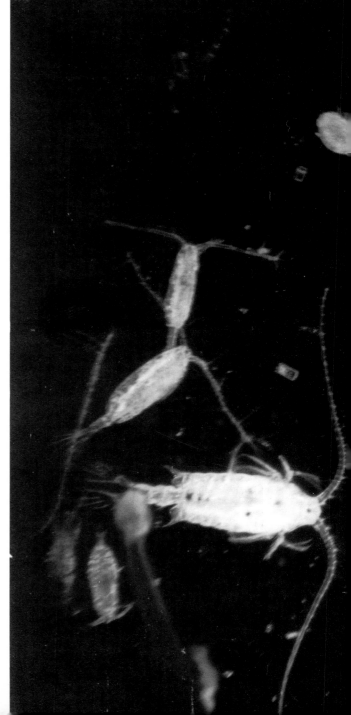

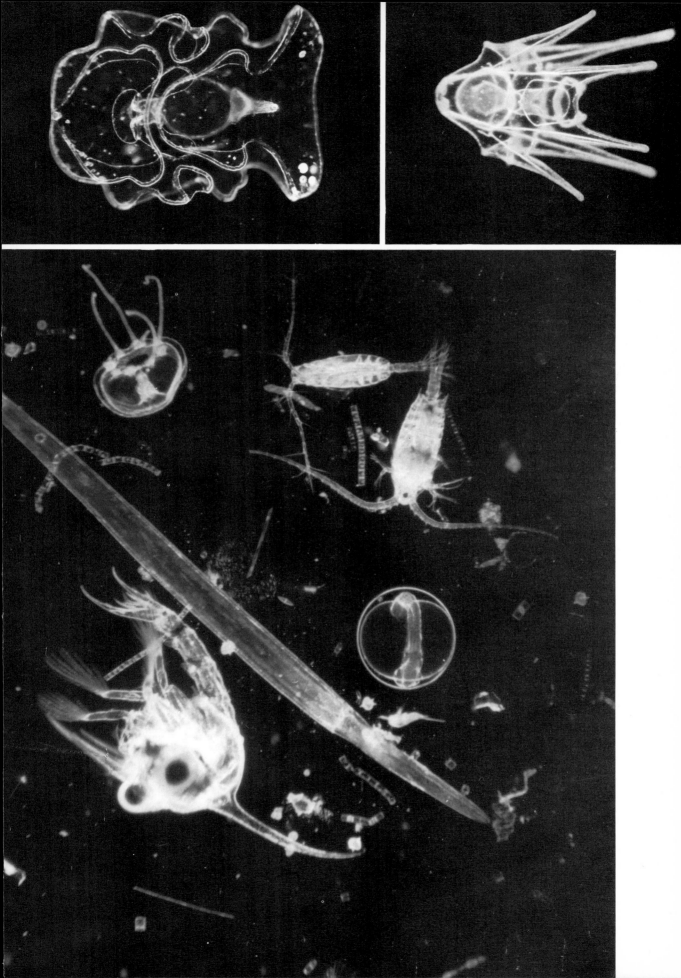

Abalones, limpets and other gastropod mollusks cling to the rocks, grazing on the green slime of smaller algae that coat the surfaces. Marine mussels (*Mytilus*) hold to the rocks with multiple threads secreted by a gland in the flexible foot; they draw in bacteria and plankton along with water for respiration. Sea anemones with saclike bodies extend soft tentacles studded with nettling cells to capture small fishes and other swimmers. Both the anemones and the mussels can move to new locations if their food supply proves inadequate. Sea squirts (tunicates), barnacles, and hydroids, by contrast, are fixed for life. Clustered together, they are normally surrounded by particles of food at high tide. While the tide is out, they cease activity. Some tolerate many consecutive hours to exposure to dry air, direct sun, temperatures that may be far higher or lower than any to be found in the sea, and carnivorous animals from the terrestrial world that do not dive for food.

The world for many barnacles is a narrow zone, bounded on the upper edge by starvation and desiccation, and on the lower by predators. The largest are fishes, which crush barnacles by the dozen and separate the flesh from the limy shells. Predatory snails, particularly the dog whelks (*Nucella* and *Nassa*), force apart the shell doors or drill through the shell to reach the barnacle. These mollusks attack mussels and other bivalves too, rasping neat circular holes through which to get at the meat inside. Some sea stars (*Asterias*) move about at a slow pace, forcing bivalve shells ajar to gain access to the living contents. These predators have no need to hurry, since their victims are anchored. By being conservative in their movements the snails and sea stars can retain a firm grip and resist being displaced by a sudden, forceful wave.

Sandy coasts offer fewer opportunities for attached plants and animals, unless certain flowering plants have gained a roothold and reinforce the bottom, or oysters have roofed it over with a thick layer of heavy shells. Along tropical coasts, mangroves (such as *Rhizophora*) grow as shrubby trees from whose outstretched branches new adventitious roots descend and resist the action of wave and wind. Oysters and many other organisms grow on the mangrove roots. Less conspicuous are turtlegrass (*Thalassia*) in warm waters and eelgrass (*Zostera*) in cold, which support diatoms and other algae on their submerged leaves, which rise by the hundreds, narrow and green but continuously immersed, from horizontal stems in the bottom. Except where tidal currents scour through regularly, these flowering plants capture sediments and humus materials, gradually making the water shallower and extending the land. Until the salt water they tolerate no longer floods in at each high tide, they offer a haven for many forms of marine life.

Hidden from sight in the sandy bottom, burrowing animals adjust their position according to the disturbances produced by ordinary waves and major storms. The burrowers include some bivalve mollusks, echinoderms (particularly heart urchins and sand dollars), worms, crustaceans, and sea anemones. Any that let themselves become exposed are likely to be eaten by sand sharks, skates, rays, or some of the larger predatory snails such as whelks and conchs.

Where the sea floor is muddy, the wave action is usually too gentle to disturb the burrowers, which find firmer sediments a few inches below the surface. But each swirl of water picks up the fine mineral particles, making the water turbid and mixing the inert material with the nourishing plankton. Filter-feeding animals must then dispose of the mineral matter, and are in danger of having their breathing organs clogged. Oysters may suspend activities for several days after a storm, until the water clears again. Yet minute plants tend to form a thin crust over the surface of the mud, nourishing sea cucumbers, some worms, and mollusks that are adapted to skim off the plant coating. Thousands of mud snails (*Nassarius*) scavenge in the superficial layer, alert for the diffusing odor of decaying flesh. Quickly they turn toward a dead animal of almost any kind.

The Frontiers of Sea Life

While the tide is high, marine organisms are buoyed up by a solution of remarkable uniformity. Despite the great quantities of soluble material brought to the seas for eons, the major solutes are stabilized at low concentrations by a whole complex of chemical interactions. Sea water remains about 96.5 per cent water. The components of the remaining 3.5 per cent, generally referred to as "salt," take the form of ions with electrical charges. Chloride ion accounts for 1.9 per cent, sodium-ion 1.0, sulfate ion 0.3, magnesium-ion 0.1. The rest are calcium-ion, bicarbonate ion, bromide ion, boric acid, strontium-ion, fluoride ion, and rarer ingredients.

The concentration of these materials rises to 4.0 per cent or more in the Mediterranean Sea, the Red Sea, the Gulf of California and a few other places because local evaporation dissipates water faster than the sea water can undergo mixing. In the Baltic and Black seas, so much fresh water enters from rivers that the concentration remains lower than that in the open ocean. These variations in the marine environment call for no special adaptations in living things that grow where the salinity continues unchanged for months or years.

At every ebb tide, however, marine life along any coast meets the harsh changes characteristic of a frontier. The living denizens in a tide pool on a rocky shore may seem safe, still surrounded by enough water to continue normal activities as though in a miniature aquarium. Yet if the sun is bright and the air hot, the water is likely to get warm and evaporate, which concentrates the salts in the residue. Or a rain shower may dilute the pool until only the most euryhaline of life in it can survive.

On a larger scale these same hazards affect shallow lagoons that are cut off from the sea by a dike of sediments left by a storm. In dry weather, a lagoon can be transformed into a salt desert, with crystals forming as the water evaporates. During a rainy period, the lagoon water becomes so dilute that virtually all of the marine life dies. The physiological stress is osmotic, for organisms can rarely maintain within their membranous coverings the inner water they need regardless of whether their outer environment is fresh or a solution saturated with salts.

Tolerance for altered salinity meets its most severe testing where the salt water of the oceans meets the fresh water from the land in the drowned river valleys known as estuaries. Where the large outflow from a river system mixes with sea water that is pulsed by the tides, the fresh water floats out over the denser salt water, carrying with it freshwater animals and plankton plants. Usually they are doomed as soon as they reach the turbulent layer along the interface, where the fresh water mixes with the salty underneath. On an incoming tide, the river water may be stalled or actually carried upstream on a strong counter-current of sea water moving along the bottom. This action enlarges the habitat temporarily for the many marine components of estuarine life, and prolongs the flushing time of the estuary.

Marine fishes and invertebrates that are stenohaline are in peril if they follow too closely the leading edge of the counter-current, since this boundary moves upstream at high tide and downstream at low. (See diagram p. 217.) To stay in full salinity they must whirl around and move seaward. Those that fail to do so are likely to be carried as corpses to a remarkable grave-yard at the mouth of the estuary. This is the region that receives also the bodies of stenohaline freshwater life that dies when whirled to the end of fresh water.

Ecologists think of the dead and decomposing matter on the bottom at the seaward end of the estuary as the catch of a "nutrient trap." This area supports a wealth of bivalves and marine worms that live in a dense community on and in the organic sediments. But much of the nourishment is spread up and down the channel by tidal movements. It reaches sedentary oysters and mussels as a dual subsidy: it arrives regularly without these filter-feeders having to expend energy in pursuit of a meal; and it consists of organic foods almost all derived originally from green plants in the river or on land. It is a subsidy of nourishment lost from freshwater ecosystems and from adjacent ocean waters, trapped, and concentrated in the estuary by the energetic interaction of flowing river and surging tides. It makes an estuary from 10 to 100 times more fertile than an equal area of open sea or of most terrestrial situations.

In temperate estuaries, animal life often shows a conspicuous programming, with different cohorts of marine animals present in each month of the year. This is equally true in the extensive saltmarshes where tidal salt water probes repeatedly up meandering gutters through stands of salt grass (or cord grass, *Spartina*). Marine frontiers of this type are common along the southern part of the Atlantic coast, especially in the Carolinas and Georgia. Each marsh is walled off to a major extent from the ocean by offshore sand bars, perhaps remaining from a line of dunes built by the wind millennia ago, when the sea level was substantially lower.

Although fresh water drains into a salt-marsh after rains inland, the salt from the sea keeps the salinity changing so much that the habitat suits few conspicuous plants other than salt grass. Wildlife in the marsh tends to be inconspicuous both winter and summer, and few animals eat the salt grass itself while it is green. The plants grow tall, then die down to the roots in autumn. Decomposers release the nutrients and also a rich detritus of organic particles. A film of diatoms coats the immersed surfaces and the bottom of each tidal gutter; surplus cells are carried away on the ebb tide. In season, consumers in the food web receive an additional benefit from the eggs, sperm, embryos, and larvae of invertebrate animals which wash up and down the gutters. So much organic matter is available that the total productivity rivals that of a rice paddy.

Fishes, crustaceans, and mollusks move into the gutters and salt ponds within the marsh, feeding on the detritus and the living plants, especially the diatoms. These are the nursery shallows for menhaden, flounders, mullet and for a great many small

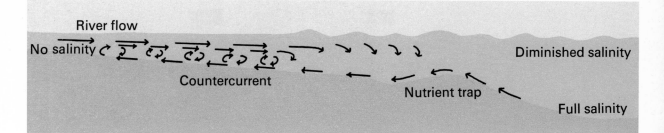

River flow

No salinity

Countercurrent

Nutrient trap

Diminished salinity

Full salinity

fishes that later spread into deeper water only to become the principal foods of cod, haddock, pollack, striped bass, and other active predators. Marine polychaete worms, crustaceans, and many mollusks (both snails and bivalves) reproduce at sea and then enter tidal waters to find preferred habitats in which to mature. They are part of the food web at the edge of the ocean, without much role in the life of the land.

Colonists from and to the Sea

Across these frontiers of the ocean world some 400 million years ago, plants and animals progressed until they colonized the land. But few succeeded with a frontal attack—up the beaches and into the air. A scattering of kinds still make this move at each generation, and compete for food and shelter with land dwellers that have a special tolerance for salt. Some of the periwinkle snails (*Littorina* in temperate and arctic regions, and *Tectarius* in the tropics) spend larval stages among the marine plankton, then creep out on shore during wet weather to browse on lichens many feet above high tide; these mollusks remain dormant, each closed within its shell by a horny door, for days or weeks of dry weather. Ghost crabs (*Ocypoda*) return momentarily to the sea to wash off their fertilized eggs where juvenile stages can swim and browse among the plankton; attaining adult form, the young come ashore and dig U- or Y-shaped burrows in the dry beach as places of concealment for the day; by night they scavenge on land.

The invasion from the sea to land for most kinds of life began as an infiltration of estuaries, saltmarshes, swamps, and rivers. The challenge of low salinity was overcome before that of drought was met. Fishes with jaws evolved in fresh waters, and so did

the amphibians, the vascular plants, and probably the bryophytes. No one knows when algae and fungi commenced their mutualistic growth as lichens, able to pioneer on land. This too may have occurred far from the sea.

A reverse invasion, by living things with terrestrial ancestors entering the marine world, has been modest. Few kinds of flowering plants tolerate immersion of their roots by salt water, as mangroves do. Fewer reproduce while exposed to a salinity of 3.5 per cent around their leaves. Even the plants that colonize the dry sea beaches take their moisture from humid air and absorb nothing through their roots. Ecologists call these plants halophytes ("salt plants") although the salt they encounter most often is in salt spray in air.

The marine world has become the only place in which certain kinds of mammals and of birds can get their natural food. Terrestrial ancestors with different nutritional requirements gave the world its whales, sea cows, and seals, all of which compete successfully with oceanic creatures that evolved through no comparable history on land. The penguins, the tubenosed birds (such as albatrosses, shearwaters, and petrels), the pelicans and their allies, the auks of various kinds, and the majority of terns are equally dependent now upon the oceans in which they catch fishes, squids, and large crustaceans that venture near the surface. To take this bounty from the seas, the arctic terns (*Sterna paradisea*) travel 20,000 miles or more each year from nest sites on land in the Far North. They cross the Equator southbound in August and northbound in late February or early March, seeing more daylight annually than does any other kind of life.

13 Life in fresh waters

Of all the water in the world, about 3 per cent contains no more than a seventh as much dissolved material as the oceans and is distinguished for this reason as being fresh. It includes the amount presently locked up in the form of ice on Antarctica and Greenland, and much that lies deep in the earth beyond man's reach. Yet this small fraction of the total water sustains the plants and animals in rivers and lakes and on land. The variety of life that depends on fresh water is far greater than is to be found in all of the oceans combined.

The organisms that inhabit fresh water rely for their habitat upon the relative constancy of the great hydrologic cycle. They rely also upon their own exceptional abilities in concentrating whatever amounts of dissolved nutrients they need from so extremely dilute a medium, and in preventing (or compensating for) an osmotic inflow of water through their membranous coverings.

The ancestors of the microbes, plants, and animals that live in fresh water came from either the sea or the land. A marine origin seems clear for the algae (both microbial and filamentous), the protozoans, the few sponges (*Spongilla*) and coelenterates, the mollusks, worms, and fishes. Some of the fishes, particularly the salmon and several kinds of trout, continue to arrive each year as fasting adults to lay their eggs in fresh water; after developing and feeding for many months in fresh water, the young of these anadromous ("upward running") fishes descend rivers to the ocean and grow to maturity there. Most of the freshwater crustaceans had remote ancestors in the sea, but some, as well as the insects and spiders, the reptiles, birds and mammals, the flowering plants (and perhaps the liverworts and mosses),

are invaders from terrestrial habitats and retain traces from this previous way of life. The amphibians probably evolved in fresh water. Virtually all of them, if they venture onto land, return to lay their eggs in ponds or streams. Some tropical frogs and toads find substitutes high above the ground in trees of the rain forest.

Rather few of the freshwater plants and animals can tolerate exposure to increased salinity. It is as though their ancestors, in adding adaptations that reduced the stress from life in such a dilute environment, committed themselves in ways that made poisonous the concentrations of sodium-ion, magnesium-ion, chloride, and sulfate that any marine organism accepts as normal. Among the noteworthy exceptions are about 28 different kinds of freshwater eels (family Anguillidae). They enter the estuaries at an early age, reach maturity in an environment of reduced salinity, and then return to the depths of the sea to mate and lay their eggs. This habit makes them catadromous ("downward running") fishes; apparently they eat nothing after leaving fresh water. This makes some fisheries men wonder whether European eels actually manage to reach the Sargasso Sea, where American eels go to reproduce. It may be that the eel larvae that continue in the Gulf Stream across the North Atlantic Ocean, and grow extra vertebrae as they go, are perpetual colonists from the American population. Although they mature and later descend to the sea in the catadromous tradition, they may never reach their final destination.

The limnologists, who investigate the freshwater world in much the same ways that oceanographers do the seas, recognize that survival in flowing waters depends upon a special set of adaptations to avoid being carried away by the current. These are quite unlike those needed in lakes and ponds where the flow is imperceptible, but stratification and stagnation introduce physical and chemical challenges that are rarely encountered in a river.

The Challenges of Flowing Waters

Anyone who lives close to a river learns how variable is the amount of water going by. In the upper reaches, each rainstorm and each day of melting snows raises the water level. The breadth of the current increases, and so does the rate of flow at midchannel. Floating objects, which may include large chunks of ice, become wrecking tools. Materials freed by erosion and carried in suspension are dropped later as gravel, sand, mud, and debris when the current slackens. By contrast, each week of dry weather lets the water level sag. A prolonged drought may divide the river into a series of pools. At this point in time, the freshwater habitat transforms abruptly from the flowing to the stationary type, and stagnation sets in. Plants and animals that were living on or in the marginal parts of the river bottom are exposed to air and perhaps sun as well.

The special adaptations shown by most life in flowing water relate to staying in place despite the current, and to surviving each cyclic period of dry weather. Fishes tend to stay close to the shore and the bottom when the water moves quickly, and to maneuver as needed while heading upstream, swimming at whatever speed is necessary to avoid being carried down by the water. Catfishes scavenge along the bottom. Trout and other game fish snap at insects and other small animals—even at mice—that are being swept along by the current.

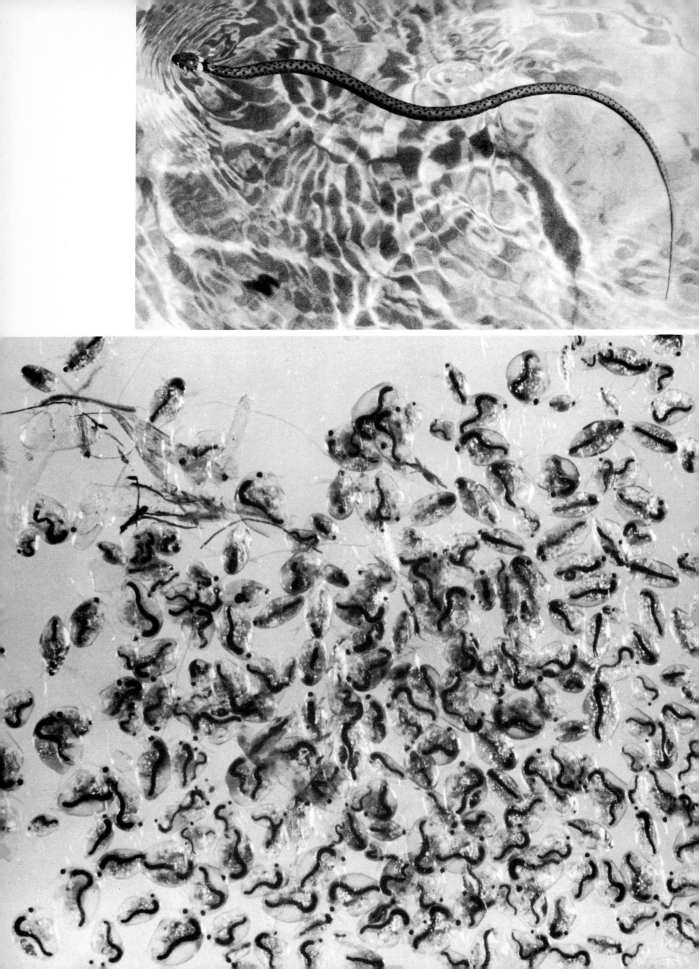

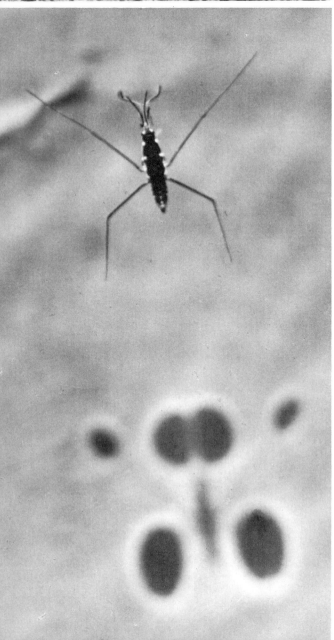

Distinctive plankton is lacking. Indeed, many animals of rivers and streams show an adaptive alternative to the planktonic larval stages that their relatives go through in marine situations. The "polyps of sweet water," as the Abbé Abraham Trembley called the freshwater hydras in the 1740's, are peculiar among coelenterates in having no medusa stage that might get lost downstream. Instead of relegating sexual reproduction to this swimming phase in the life history, the hydra polyps develop testes or ovaries or both. The ciliated larva that emerges from the fertilized egg quickly settles to the bottom and transforms into another polyp.

Fresh waters do have medusae (Craspedacusta is virtually worldwide) as much as $\frac{3}{4}$ inch in diameter, with a diminutive polyp stage no more than $\frac{1}{8}$ inch tall. These medusae were discovered first in a display tank full of giant water lilies in Kew Gardens, England. Since then they have turned up in slow rivers and in small lakes— even in private swimming pools! Apparently the polyp stage can be carried in moist mud on the feet of wading birds, and travel where they do.

Bivalved mollusks, which remain partly buried in the sediments on the river bottom, are adapted to use fish in keeping their larvae in fresh water. Unlike marine bivalves, they risk only their sperms into the aquatic environment. The female mollusk retains her eggs in her pouchlike gills. After the eggs hatch, their parent ejects them a few at a time through her excurrent siphon into the river water. There the hatchling, called a glochidium, begins to snap its tiny valves, each with a tonglike projection, as though biting at the water. By chance it may touch the body of a fish or be engulfed by one. Quickly the glochidium attaches itself to the skin of the fish or to a filament

221

on the gills. It then rides along as a parasite, continuing its growth to the critical stage at which it can drop off and transform into a miniature facsimile of its parents. Thereafter it will feed on suspended bacteria and other organic material in the manner so characteristic of bivalves.

Freshwater snails are equally adapted to living in a liquid environment that moves toward the sea. Lacking planktonic stages, they develop into adult form before emerging to fend for themselves. Some do so in the eggs that the parent snail affixes to a solid object under water. Others are born in an active condition by a member of an ovoviviparous genus (*Viviparus* and *Campeloma*).

Diatoms, liverworts, and water mosses that grow on stones in rapids and riffles provide food for the snails and the larval stages of many insects. These animals avoid the current, rather than resist it, by having a low silhouette and staying so close to the rock that they are in a layer of water slowed greatly by friction. Or they cling to emergent plants, such as the reeds (*Phragmites*) along temperate rivers and the papyrus sedge (*Cyperus papyrus*) along the Nile, which grow where they have moisture during the driest season yet, by being close to shore, not too much erosive action during floods.

Juvenile stages of insects in many orders inhabit running water: mayflies (Ephemeroptera), stoneflies (Plecoptera), dobson flies, alder flies, and spongilla flies (all Neuroptera), net-building and case-building caddisworms (Trichoptera), some midge and blackfly larvae (Diptera), and most conspicuously the "water penny" larvae and adults of parnid beetles (Coleoptera). Most of these complete their aquatic life while water and food are plentiful, then

Amphibious Adaptation
The roots of the tropical papyrus reed grow in water-soaked soil along a river or pond, while the leaves and distinctive stalks with flowers or fruits rise tall into the air (above). Tiger salamanders (above, right) may crawl out on the muddy banks of the pond, or swim lazily in the water, much as their four-legged tadpoles do until each develops lungs to replace its shrinking gills. Whole forests of bald cypress trees (right) grow in southern wetlands, getting oxygen to their submerged roots through the porous tissue of "knees" that extend above the water.

222

bridge the dry season and the winter in the pupal stage, or as adults hiding on land, or as eggs that will not hatch until the spring floods are past.

The immature mayflies and the "blood worms" that are the larvae of certain chironomid midges prove to be ideal indicators of water quality. Where the water is well aerated, such as below a rapids or a waterfall, the young mayflies thrive. They die for lack of oxygen where decomposers compete with them while working on organic pollutants, and may be absent altogether for several miles downstream from the point where the pollution enters. Where almost no oxygen remains, the blood worms thrive, for they can feed anaerobically on detritus and in such places have few controls to limit full expression of their biotic potential.

Standing Waters

By the end of summer, blood worms are often the only animals among the detritus at the bottom of a lake more than 30 feet deep. Since spring, when all of the water in the lake had the same temperature and maximum density, the bottom water has been isolated from circulation and become progressively more stagnant. Limnologists call this water the hypolimnion, and know that it will be virtually unproductive until the lake is again uniform in temperature and density from top to bottom. Then the wind can set it all into rotary motion for an autumn overturn that is the counterpart of the one in spring.

Standing waters differ greatly in the amount of dissolved nutrients available to green plants. Mountain lakes, fed from melting snows and glacier ice, sometimes contain so little phosphate and nitrate that they are scarcely more hospitable to life than distilled water. Ecologists refer to them as oligotrophic ("barely nourishing") and know that they have plenty of dissolved oxygen all the way to the bottom because almost no life is there to carry on respiration. If inorganic fertilizer, such as farmers use for plant crops, is added to the waters of such a lake, its productivity increases suddenly. The change, called eutrophication, transforms the desolate body of water into one that is more like the lakes that receive an infusion of soluble mineral matter each year by natural processes. These lakes are naturally eutrophic, and begin each annual growing season with enough dissolved nutrients and oxygen to support a rapid growth of plants and animals. By midsummer, however, the growth of floating and drifting green plants in the epilimnion of a eutrophic lake tends to cease for lack of more dissolved phosphates and nitrates to incorporate into new biomass.

Eutrophication of streams and lakes with phosphates and nitrates has become a recognized form of environmental pollution where farm fertilizers wash into fresh waters from crop lands or sewage from communities brings the decomposition products from modern biodegradable detergents. These additional nutrients arrive so steadily that natural processes become disastrously overloaded. Early in the annual growing season, the phytoplankton and the bottom vegetation are stimulated to tremendous productivity. In surface waters the plants become so dense that they block the light from reaching moderate depths. Deeper plants die. Under the green covering, decomposers in the epilimnion and mesolimnion use up the oxygen, suffocating both the fishes and the animals that fishes eat.

Even this deterioration of the freshwater

environment can be tolerated by many of the snails and of the insects in standing waters. For oxygen, they come to the water surface and break through the surface film to inhale air. With a bubble captured into a lunglike sac within its mantle, the snail descends to feed for a while. With air under its hard wing covers (elytra) or taken directly into its respiratory tubes, the insect (a beetle, a mosquito wriggler, or a water bug) dives and resumes its aquatic life. A "rat-tailed maggot" (the larva of a drone fly, *Eristalis*) can scavenge among decaying organic matter while carrying on gas exchange through a telescoping projection like a snorkel from the rear end of its body. The tube is as much as six inches long on a one-inch maggot.

A few kinds of insects are even more remarkably adapted, for they swim half-exposed to air (as whirligig beetles do) or stand dry-shod upon the water film. Whirligigs scavenge at the surface, swimming in characteristic zigzag patterns, often in groups and at amazing speed, or they dive briefly for additional food or for safety. Long-legged water striders patrol for flying insects that fall in and drown or that die below the surface and float upward to the boundary of the wet world. Springtails (order Collembola) devour such inconsequential foods as pollen grains that fall atop the water, before the particles become wetted.

Temporary Lakes and Ponds

Many a shallow pond or small lake dries up each summer, long before any clogging coat of vegetation spreads from shore to shore. It may still harbor a succession of small animals, whose cycle of development fits into the short annual schedule. This program is perfect for mosquitoes, which need only a few weeks from egg to adult stage. It fits the program inherited by fairy shrimp (*Eubranchipus*) which hatch from eggs that have laid dormant from late in one spring to early in the next. It is adequate for a whole series of different frogs and toads whose tadpoles will grow legs and shrink their tails to achieve adult form before the water vanishes. Salamanders join the parade to the ponds, to engage in mating ballets and deposit their fertile eggs. Salamander larvae hatch with good jaws and four legs, ready to pounce on insects, small crustaceans and worms, whereas tadpoles at least begin life legless and as vegetarians with sucking mouths.

The most temporary of all bodies of fresh water are those that fill low places in the desert after a sudden rain. Several years may has passed since the particular lake or pond held water. But among the dust are dormant, living cells of algae and protozoans. Within two days these organisms are reproducing at a prodigious rate. The slightly larger but equally desiccation-resistant eggs of brine shrimp (*Artemia*) and tadpole shrimp (*Triops*) hatch into tiny crustaceans that grow quickly on an algal diet. Snails that have been hiding deep in the desert soil soon work their way to the surface and browse on the same kinds of plants.

Dangerously Hot Waters

Most kinds of life die from prolonged immersion in water at a temperature much above 100°F (40°C) because their proteins are inactivated, eliminating essential enzyme systems. Yet in those parts of the world where aquatic habitats at higher temperatures have persisted for thousands or millions of years, a number of species have evolved the tolerances needed to live and reproduce in dangerously hot waters. Preventing heat death may have been easier to evolve, through natural selection of adaptive rearrangements at the molecular level, than ways to overcome other limitations in the environment of thermal springs. Where the water is heated by hot gases escaping from volcanic vents, it is often charged with carbon dioxide. Surface waters that penetrate far enough underground to be heated close to the boiling point by rocks still hot from recent volcanic activity may return to the surface rich in sulfur- and calcium-containing solutes. Upon cooling, they commonly deposit terraces of many shapes. Rarely does the water contain any nitrogenous nutrients.

The living conditions suit some of the autotrophic sulfur bacteria and the blue-green algae that fix their own nitrogen. Where the water has cooled a few degrees, herbivorous larvae of certain flies may feed on the bacterial coating and the algal mats. The other common members of the limited biota are mostly predatory larvae of other flies and parasitic mites. Decomposers, gaining nourishment from dead algae, arthropod feces and cast skins, and dead bodies, may be few because the wastes and corpses progress downstream. There, in cooler waters that challenge life less severely, the nutrients are in greater variety. They support conventional microbes and familiar plants and animals.

Poisoned Waters

Around the edges of cold lakes, and on ill-drained mountain sides and tops, the peat moss (*Sphagnum*) colonizes and proliferates, changing the conditions of the habitat in ways that sharply diminish the diversity of flora and fauna. The leaflike branches of the moss contain living cells with chlorophyll, and also many empty cells with stiffened walls and a pore to the outside world. So long as these empty cells contain air, they buoy up the plant and let it float out over water. But as the moss grows, it shades the lower older parts until they die. Partial decomposition lets water replace the air, and the weight of overlying living moss pushes down the nonbuoyant material. Exudates from it are rich in acids and tannins, which inhibit the activity of decay bacteria. Most larger plants and animals find the acids and tannins intolerable. The lake becomes a bog as the moss roofs it over with a quaking mass. Ecologists refer to the poisoned water as dystrophic, since the dissolved substances limit the proliferation of life instead of supporting it.

A number of woody plants take root in the moss. Most conspicuous are larch (or tamarack, *Larix*) and spruce (black spruce, *Picea mariana*, in the New World, and Siberian spruce, *P. obovata*, in the Old). These and the low-growing shrubby members of the heath family (Ericaceae) seem shielded from the bog water by the fungal partners (mycorrhizae) that surround their roots. An arctic sedge called cottongrass (*Eriophorum*) and various insectivorous plants grow well. Each uses its roots only for anchorage, absorbs what little moisture it needs through its foliage, and limits these needs by adaptive features of leaf and stem that make it a true xerophyte—a drought plant—living with its roots in water. The drought is physiological, yet chronic, shutting off dissolved phosphates, nitrates, and other mineral nutrients. The insectivorous plants get these materials from the bodies of the insects they catch and digest.

Few pioneers from surrounding land can challenge the special plants of the bog community. It perpetuates itself until the dystrophic water drains away through some channel opened by new movements in the earth, or is displaced by solid peat.

In the northern half of North America and of Eurasia, peat bogs have maintained themselves in the same locations since the glaciers of the Ice Age retreated, leaving a pattern of scoured rocks and debris with poor drainage. Into the cold waters, the wind dropped pollen grains and other vegetable particles representing the plants near by. The grains sank, did not greatly decompose, and accumulated in subfossil form. Recently, scientists specializing in pollen identification have examined in detail long vertical core samples cut from ancient bogs. Their identifications allow a reconstruction of plant succession corresponding to weather changes since part way through the Pleistocene—the years in which the human species spread and became aware of the environment.

Wet Edges of the Land

Most wetlands form a temporary edge between deeper, open, fresh water, and land that has a soil of some kind. Customarily, only decades (rather than centuries) pass before the natural succession among living things transforms the wet shallows into either a forest tract or a grassland, depending upon the climate. Where the rainfall is well distributed through the year, trees emerge from the shallows and produce a swamp. Where the annual precipitation is limited and least in summer, trees are rare and monocotyledonous plants with coarse grasslike leaves grow up, producing a marsh. Silt and undecomposed organic matter accumulate in both swamp and marsh until the low, water-filled areas are obliterated.

In a few areas of the world, the characteristic succession has been stalled for centuries or millennia in the marsh or swamp condition, mostly because the water spreads over an immense territory that is almost level and the current is too slight to transport a load of silt. This is true of the vast marsh in the southern Sudan where, in the midst of deserts, the White Nile seems lost in more than 50,000 square miles of flatland called the sudd. Through this roadless marsh a navigable channel is maintained with difficulty because great rafts of papyrus stems drift about, supporting an aquatic grass (*Vossia*) and some thorny trees (*Herminiera*) of especially rapid growth. Water lettuce (*Pistia*), which fish fanciers often keep as floating greenery in their aquarium tanks, is native. Water hyacinth (*Eichornia*) was introduced in 1957, further clogging the channels through the sudd.

The Everglades in southern Florida is another "sea of grass," dominated by a coarse sedge known as sawgrass (*Cladium jamaicense*). The area extends about 70 miles east to west and 40 miles in the opposite direction. It is studded with small islands bearing clumps of cabbage palm and saw-palmetto.

Dense and impenetrable as these regions are for people on foot, their productivity is modest. Even where the weather is generally warm and the low latitude corresponds to a high sun each day, the abundant water is deficient in dissolved mineral matter. It differs from the medium in oligotrophic lakes in containing a considerable load of organic compounds. These color the water, but offer less energy to decomposers than might be expected. Neither a freshwater marsh nor swamp can compete with a saltmarsh except as a refuge for birds that find fewer amenities elsewhere.

14 The hidden life of the soil

Until well into the twentieth century, the role of living things in producing and renewing the soil remained widely unappreciated. The fact that microorganisms completed the conversion of organic wastes into useful substances went virtually unnoticed. Attention was focused almost exclusively upon the parent materials such as pulverized rock particles of various origins, sizes, and chemical nature. Various physical and chemical processes appeared to produce and amass the mineral particles. The disintegration of rocks provided the materials that were carried to the site by rock falls, landslides, mud flows, running water, glaciers, or wind. The eminent German chemist Justus von Liebig promulgated this view of soil as inert and essentially lifeless in his writings of the 1840's. To him fertile soil differed from sterile soil only in the amounts of inorganic nutrients—such as lime, potash, nitrates, and phosphates—dissolved in the water within reach of roots. He wrote that "crops on a field diminish or increase in exact proportion to the diminution or increase of the mineral substances conveyed to it in manure." Many people still think of the soil in this way, particularly when they buy bags of fertilizer. They seek to meet the needs of lawn grasses by choosing, for example, prepared mixtures labelled 10-10-10, indicating 10 parts of lime, 10 of potash, and 10 of nitrates, plus 70 parts of inert matter.

In retrospect, it is easy to find isolated works with a more dynamic view of soils. In 1860, E. W. Hilgard's *Report on the Geology and Agriculture of the State of Mississippi* stressed the interaction of climate and particular kinds of vegetation in producing soils from eroded rock materials. The experimental studies that Charles Darwin conducted and described in his book *The Formation of Vegetable Mould, through the Action of Worms, with Observations on their Habits* (1881) could have stimulated a broader understanding of the soil as a community. But it was not until after 1870 that a number of Russian scientists, under the leadership of V. V. Dokuchaev, discovered that soils have a structure, produced by specific processes. Language barred any effective spread of this concept to western Europe and America until 1914, when K. D. Glinka, a Russian, summarized it in a book in German. In 1927 it was translated into English as *The Great Soil Groups of the World and their Development*. Some of the Russian words for soil types and processes have been adopted almost universally.

Glinka pointed out that where moisture reaches land with vegetation on it, the mineral particles (called parent materials) in the soil are modified and develop a structure. History is recorded in the succession of layers that can be seen in a vertical section (called a profile) down into loose unweathered rock. Dead leaves and stems, animal bodies and droppings that are still recognizable lie atop the soil and are sometimes referred to as the O horizon. Below them is the topsoil, or A horizon, in which mineral particles are intermixed with the organic residues from the litter. Ordinarily, the upper portion of the A horizon is dark and known as humus, whereas the lower portion has lost some of its color due to removal of soluble components by water sinking toward underground drainage channels (called aquifers). Most decomposition occurs in the A horizon, and it is there the greatest number of living things are found.

Between the A horizon and the C, which consists of essentially unmodified mineral particles and continues downward to bedrock, there may be a recognizable subsoil or B horizon. Processes known collectively as mineralization give this layer a blocky or prismatic structure quite unlike that of either the A or C. Generally it is firmer when dry, stickier when wet, and less attractive to living organisms except where roots penetrate it. The B horizon contains many remains of dead roots and fragments of plants that earthworms and small mammals have dragged deep into their burrows. It does, however, ordinarily contain clay and soluble inorganic substances carried there by water sinking down from overlying layers.

In the tropics, where the annual precipitation is great and the soil supports a rain forest, the A horizon is virtually absent because decomposition is so rapid. As fast as materials are made soluble, they are dissolved and either washed away over the soil in run-off water or carried downward by leaching into underground streams and eventually to the sea. Left behind is a reddish yellow mud, consisting of virtually insoluble sesquioxides (sesqui = $1\frac{1}{2}$) of iron (Fe_2O_3) and aluminum (Al_2O_3), and granules of yellow limonite ($2Fe_2O_3.3H_2O$), forming a thick B horizon. Glinka called this process laterization and the soil a latosol, from the Latin word for brick.

In a temperate or cool climate with only a moderate amount of precipitation leaching is less severe and decomposition slower. Organic materials accumulate into an A horizon. If the growing season is long enough, the land will support a forest dominated by deciduous trees. A shorter season favors a mixture of deciduous and evergreen trees, or conifers almost alone, or shrubby heaths. The fallen branches and leaves from these kinds of vegetation differ chemically from those of the rain forest and provide a feedback to the soil that affects the processes of soil formation in important ways. Most deciduous trees are relatively low in lignins and tannins and relatively rich in compounds of calcium and magnesium. Products of their decay neutralize organic acids and carbonic acid in the soil, leaving it slightly alkaline. This state is favorable for the decomposers and also the germination of many kinds of seedlings. It is a positive feedback to the soil, maintaining a fairly thick A horizon and a nutrient-rich B horizon. Generally the soil is brown.

Shorter growing seasons and cooler climate at higher latitudes match the adaptations of conifers and heaths. Under these a litter accumulates that is rich in lignins and tannins, often in resins too, but poor in compounds of calcium and magnesium. Decomposition of the lignins, tannins, and resins is slow, and produces acids that are not neutralized. Instead, the acidified soil moisture attacks the sesquioxides, making clay and other colloidal substances mobile. They are leached at least into the B horizon, leaving an ash-gray layer of silica (sand) below the organic matter of the A horizon. The soil is called a podzol from the Russian word for this color. The process that produces it is podzolization.

Between the tropics and the northern coniferous forests, the soils below woodlands of the temperate zone show varying combinations of laterization and podzolization. They are referred to as podzolic soils and found to correspond in fertility to the amounts of calcium and phosphorus compounds present. These are low in oak forests because oak leaves contain so much tannin and decompose so slowly. They are low too wherever repeated brush fires destroy the organic matter before it can

decompose part way and slow the loss of nutrients by leaching. The ashes from the fires are quickly washed free of soluble inorganic matter.

In natural grasslands and savanna country, the annual precipitation does not suffice to leach away soluble materials. The O horizon is very thin because the accumulation of dead leaves and stems is so much less than in a forest. But the A horizon is generally thick because grass roots die and decompose in 5 to 7 years, as compared to many decades for tree roots. Decomposers work on the grass roots all summer, releas-

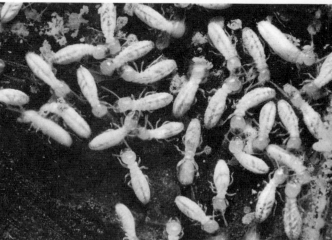

ing nutrient materials for new growth in a soil that is neither strongly acid nor alkaline. Rain that sinks in carries the sulfates and carbonates of calcium (and magnesium) down through the B horizon. There the deep roots absorb the moisture, letting the salts crystallize out as a layer referred to as the hardpan. Logically, the process is called calcification. In regions where the total annual precipitation is small, the hardpan may be only a foot or two beneath the soil surface. It limits the downward penetration of both water and roots, producing what is known as a hanging water table. In rainy weather the plants suffer from excess water around their roots. As soon as conditions become fairly normal, much of the water evaporates and lets the plants suffer from drought.

The deep soils of grasslands where precipitation is adequate for agriculture develop a dark brown or black color from organic matter and are called chernozems. Land with less natural moisture has chestnut to brown soils that can support a good stand of grass for grazers, but needs irrigation during all except wet phases of the climatic cycle if it is to be used for profitable agriculture. These soils grade off into semidesert and desert wherever the available moisture during the growing season is significantly less than the potential evaporative losses.

In the tundras of the Far North, where the mineral particles of the C horizon are held firmly in a matrix of ice that has not thawed since the Ice Age, this permafrost is overlaid by a blue-gray or greenish B horizon and a peaty A horizon. The color is due to poor aeration and the resulting anaerobic conditions which favor the existence of iron compounds with a minimum content of oxygen. Low temperature retards decom-

position, the activity of soil organisms, and the spread of roots. Although the low sun melts the snow cover and thaws out the A and B horizons, drainage is generally poor. Soil moisture is made acidic by decomposition products of peat moss (*Sphagnum*) and heaths.

During the short arctic summer, physico-chemical and biological changes produce a sticky, compact, structureless mass called a glei at the bottom of the B horizon. The complex process is known as gleization. When autumn chill produces a frozen roof over the B horizon and compresses the glei above the permafrost, the material oozes upward through cracks and weak places, leaving the surface marked visibly in polygonal patterns that are quite evident from the air. Known as "frost boils," these events provide almost the only mixing of the soil. Each polygon is ordinarily bounded by narrow trenches filled with peat moss.

Mountain slopes have similar vegetation but rarely develop a glei layer. Usually they bear patches of stony soil, cupped in ill-drained cavities of the underlying rock. Dystrophic bogs are common. Drainage leads to podzolization.

Far less structure develops in desert soils, except along intermittent streams where shrubs and trees with deep roots can reach moisture most of the time. Because the available water is scanty in comparison to the evaporative power of the air, rain water that sinks in is likely to rise again by capillary action and be evaporated from the surface. It leaves in the uppermost layers whatever materials it has dissolved, producing a saline crust. Glinka called this process salination.

The mineral matter deposited along the banks of great rivers is called alluvial soil. It shows little real structure, although it may be layered in a pattern that records one flood after another. Each brings a new deposit of silt and a random array of organic matter, much of which soon decomposes. Living things on alluvial soil must be adapted either to escape the floods or to survive submergence while swirling waters bear heavy floating objects as battering tools.

Glinka's processes of soil weathering (the top term in each of the two groupings) thus match both the climate and the type of dominant vegetation (in italics):

Soil-forming processes and soil groups of the world

Climate	Warm	Temperate	Cool	Cold
Moist	Laterization Latosols *Rain forest*	Laterization and Podzolization Brown podzolics *Deciduous or mixed forest*	Podzolization Podzols *Taiga*	
Semiarid to Arid	Salination Red Desert Soils *Desert*	Salination Gray desert soils *Desert*	Calcification Chernozems or Chestnut soils *Grassland or Savanna*	Gleization Tundra soil *Tundra*

Soils have become a focus for attention by scientists with diverse interests. Those who consider the subject from a geological background, as mineral materials modified by weather and living things, call themselves pedologists and their field pedology. Edaphology, by contrast, is the study of the soil (or edaphic) factors that affect the welfare of plants and animals; it is a branch of ecology. On the other hand, agronomists seek to use a scientific understanding of soil structure and edaphic factors in plant growth to devise methods for managing agricultural soils to improve crop yield.

Soil as an Environment

For plants and animals that are small enough to find a living space within the soil, conditions of existence are far more uniform than above the soil surface. Daily changes in temperature decrease with depth until, at about one foot down, they disappear entirely. At five feet underground, the temperature remains the same within a degree or two, winter and summer.

Soil moisture is essential for the activity of most soil dwellers. It is renewed by heavy rains and from melting snow, which percolate through channels left by rotting roots and burrowing animals. The extent of these open spaces is often amazing, amounting in forest soils to as much as 60 per cent of the total volume. Tunnels made by earthworms, wireworms (clickbeetle larvae), cicada nymphs, and moles often extend for considerable distances. The underground galleries of ants are less often shared, for the ants patrol them efficiently and destroy most invaders.

At times, water displaces so much of the air in the soil spaces that respiration becomes difficult for roots, aerobic bacteria and fungi, and soil animals. More usually a film of moisture coats the soil particles and serves as a microhabitat for aquatic life underground: unicellular algae near the surface, bacteria, protozoans, rotifers, and roundworms deeper down. Under most circumstances, the relative humidity of the confined air is close to saturation, which makes the spaces between the soil particles attractive to terrestrial animals that have scant protection against desiccation: mites, springtails (Collembola), and minute land snails. High in the O and A horizons, larger animals find bigger spaces and more frequent drought. Millipedes, centipedes, and sow bugs (isopod crustaceans) cannot encyst, as do so many of the aquatic organisms in soil, but must save their lives if possible by burrowing more deeply.

In soil, light penetrates only shallowly and wind is altogether absent. But the downward flow of dissolved inorganic material and decomposable organic matter is far larger and faster than is ordinarily supposed. Rain water that has washed across the surface of a living leaf or drained down a stem or over bark is no longer pure water. It is "leaf wash" and "bark wash," containing measurable amounts of potassium-ion and calcium-ion, of chloride, nitrate, sulfate, and other ions too. Dripping to the soil, the dilute solution may either sink in and be held where these nutrients can enter roots for a new cycle through vegetation, or be lost in runoff water and into underground streams.

The Belgian ecologist Paul Duvigneaud calculated from measurements made in a forest he was studying that the annual addition of litter to the soil amounted to 4 tons per acre. This same area included about 2 tons of roots and 0.3 ton of nonvascular vegetation, most of it engaged in

Moles, although scarcely larger than mice, are strong members of the mammalian order Insectivora. They tunnel through the soil, forcibly bulging it upward into characteristic mounds (top) while hunting for earthworms and insects. At night, moles often emerge to find their prey (bottom) but make so many rustling sounds that a prowling cat or an alert owl has a good chance to catch one.

decomposition processes. On the average, each ounce of the forest soil contained some 2.7 billion bacteria, 1.0 billion actinomycete fungi (some of them serving as mycorrhizae), and 28 billion other fungus plants. This living and nonliving organic matter nourished approximately 1.0 ton of soil animals per acre, of which about 0.6 ton were earthworms. From collections made with specially-designed instruments, Duvigneaud estimated that the lesser denizens included 25 to 300 billion nematodes per acre, 4 to 10 billion mites, 400 to 2,000 million springtails, 40 million immature insects, 15 million proturans, 1.5 to 12.5 million centipedes and millipedes, and astronomical numbers of protozoans.

The organic compounds in various stages of decomposition add importantly to the colloidal matter in the soil (that is, particles small enough to remain suspended in a fluid), which otherwise consists chiefly of clay particles. Colloids retain water where roots can reach it, and adsorb the ions of inorganic substances, reducing loss by leaching. Fertility of the soil is thereby increased, since roots can later remove the adsorbed ions selectively as the conspicuous plants need the nutrients for growth and other processes.

Just as colloids show physical properties that make them a fourth state of matter, different from any solid or liquid or gas, so too the majority of soil organisms differ markedly from those exposed in aquatic or other terrestrial situations. Some biologists speak of these hidden creatures as cryptophytes and cryptozoans. The differences arise from small size and a relatively huge surface area in proportion to volume. In colloidal particles, the distinctive features are produced by an electrical charge borne on the surface. In soil organisms, the disproportionately large surface eliminates any need for breathing organs but also introduces a special vulnerability to desiccation.

Among the arthropod animals of the soil, members of primitive orders predominate. It seems possible that their ancestors moved directly from aquatic habitats to spaces in the soil, and that the uniform conditions found there have stimulated little subsequent evolution. Depending upon its original structure, an animal protected from sun and wind, in uniform temperature at high humidity, could add adaptations related to the acquisition of food and thereby refine its niche. In this way, presumably, the dramas in miniature beneath our feet had their beginning: the cryptophytes and roots making available a smorgasbord of organic compounds, containing energy captured by green plants in sunlight, and the cryptozoans eating from the plants and one another.

Some of the cryptozoans—chiefly those that attack the roots of crop plants or that are active predators—have become well known. The immature cicadas live on plant juices far below the surface for as much as 17 years before emerging as adults. The aphids (plant lice) suck from rootlets; each insect remains where it was placed by an ant and is "milked" at intervals by solicitous ants that lap up the excess sugary solutions it excretes. The myriad scavengers serve as prey for predators—centipedes with poison claws, and roundworms with triple jaws.

The variety, the numbers and the total biomass of the living community in the soil all vary according to the climate and, to some extent, the chemical nature of the parent materials in soil. A. G. Tansley, who coined the word ecosystem in 1935, expressed these relationships in a diagram with six avenues for influence and feedback among four factors:

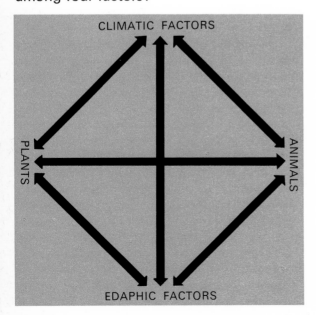

The interaction between soil (edaphic factors) and plants is primarily one of supplying mineral nutrients on the one hand and of receiving decomposable products and being opened up by roots on the other. Similarly soil (or soil water that bubbles from springs into streams) provides specific mineral substances to animals, such as the salt at salt licks and the lime needed for production of skeletons and teeth; and the soil receives from the animals their feces and dead bodies, upon which decomposers work. Soil affects climate according to its reflection of solar energy and its absorption of precipitation, and is affected by variations in temperature and moisture, which alter the rate of decomposition and the movement of materials in solution.

The climate, plants, and animals produce a soil from whatever parent materials are available, and in this sense are dominants in these relationships. Climate may allow deciduous trees to grow, and beavers to cut the trees for food and open up the forest along ponds and streams. It imposes many restrictions on warm-blooded animals and regulates the activity of cold-blooded ones. Each type of climate has its matching types of vegetation in terrestrial ecosystems known as biomes, each demarcated by a measure of evaporative stress reflecting both precipitation and temperature. The vegetation, in turn, greatly modifies the climate close to the soil. Animals have less obvious effect on climate.

Since the major soil types and biomes—tundra, taiga, temperate deciduous forest, grassland, desert, savanna, and tropical rain forest—correspond and can be marked on a map, they are features of biogeography. Yet their boundaries follow no simple combination of annual means for precipita-

In tropical biomes, the physical environment for living things changes with the seasons and geographic directions far less than in temperate and polar zones. Average temperature and evaporative rate vary chiefly with altitude, and the climax vegetation follows a pattern that reflects the degree to which precipitation exceeds or falls short of evaporation of water.

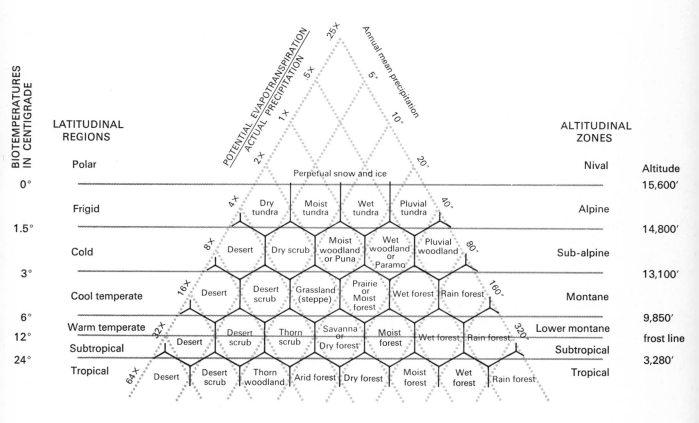

tion and temperature. A mean temperature value is more realistic for plant growth if it is calculated by adding together only the monthly means in temperature for those months during which the weather stays above the freezing point.

A plant ecologist with extensive experience in the American tropics, L. R. Holdridge, tried in 1947 to construct a diagram with predictive value, in which temperature during the growing season formed a series of levels corresponding to altitude. Rather than construct a rectangular grid by using precipitation as ordinates, he inclined these gradations and crossed them to make a hexagonal pattern with another measure, showing the moisture stress on plants—the extent to which potential evaporation exceeded or was less than actual precipitation. The outcome (see above) approximates a pyramid capped by perpetual snow, with tropical desert in one lower corner and the wettest of tropical rain forests in the other. These relationships hold well between the tropics, where the high angle at which sunlight strikes the soil changes little through the year, and plants need never be fully dormant because winter months of

By plotting the average precipitation and average temperature for each month of the year and joining these points to form a closed polygon, a climatogram is constructed as a graphic means for comparing these environmental conditions in different localities.

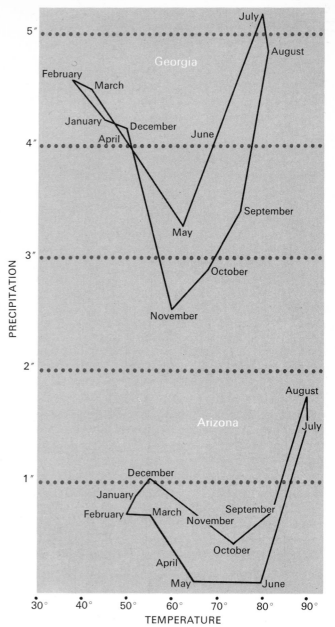

cold weather have locked up the water supply. Still another axis would be needed to make the diagram apply to biomes between the tropics and the poles. It would give a measure of growing seasons—the number of months annually with no appreciable frost—and also the direction of slopes, whether equatorward or poleward, and hence the daily duration of sunlight in lands where the source of life's energy is never overhead.

An alternative, and less quantitative, representation is often preferred for comparing two or a few ecological communities in temperate and polar regions. Called a climatogram, it shows matching values of average temperature and precipitation month after month on a rectangular graph, generally with initial letters for the months added to allow a reader to follow the changes around the cycle of a year. This mode of analysis has helped move the scientific study of ecological associations beyond the descriptive stage, replacing the question, "What life is present?" with a better one: "Why do these organisms live where they do?" Yet the "why" question could not precede or accompany the "what" until the nature of the interactions among plants, animals, soil, and climate had been studied in detail for each biome.

Cavernicolous Life

In regions where the soil is underlayed by thick layers of limestone, the cutoff water that sinks into subterranean drainage channels generally keeps dissolving the rock and opening series of connected caves. Through cracks and crevices, the runoff water from streams and rivers may enter and enlarge these underground spaces, particularly at flood season. Often the roots of trees penetrate some open chambers and weaken

the roof. Eventually it falls in, creating a sinkhole and a place for easy access to the cave system for small animals. If the caverns develop openings to the outside world at different levels, they become well ventilated by the downward movement of cool dense air. Yet the temperature may vary only a fraction of a degree all year. The relative humidity is often maintained at a high value by evaporation of moisture from walls, roof, and floor.

If useful energy is available in the cave system, it becomes suitable as a habitat for life. Living tree roots that penetrate afford nourishment to insects with sucking mouthparts. Ants may tunnel downward and milk the aphids (plant lice) of honeydew. More often the nourishment for scavengers and decomposers in caves comes from overlying soils in the form of organic matter in the water that drips from the roof and walls. Fungi thrive on it, and on many substances left in silt as the stream through the cave system spreads and contracts with variations in precipitation in the world above. The stream may bring plankton that is still alive, although it cannot survive long in the constant darkness. Cave salamanders, cave fishes, cave crayfishes, and cave turbellarian flatworms benefit from these foods in the wet parts of the caverns. But most commonly, the nourishment for cave dwellers takes the form of guano, dropped by bats and other animals that feed outside the cave and then return regularly to spend the daylight hours. The guano is so rich in nitrogenous materials that in many caves it has been mined for fertilizer or for making gunpowder. On bat guano, molds grow well. Cave crickets feed on the mold. Cave centipedes and cave spiders catch the crickets. Ectoparasitic insects cling to the bat roosts, waiting for the return of their hosts with a fresh supply of energy.

Unlike the denizens of the lightless depths of the sea, none of the cave animals are luminous. Most, in fact, are blind or have degenerate eyes. Neither blindness nor pale color is a liability in the darkness, as either would likely be in a lighted habitat where predators have keen vision. The cave animals show compensatory reliance on long sensory feelers, or smell, or taste. Many of these features are to be found, like preadaptations, in their nearest kin. These usually live in or on the soil near by, supposedly as the stock from which cavernicolous forms evolved.

15 Life in the forests

When Joyce Kilmer's famous poem *Trees* appeared, its message, "only God can make a tree," found favor among English-speaking people over much of the world. In them it appealed to an age-old respect, which is almost peculiar to Western civilization, for vegetation that grows tall, sturdy, and old, lending shade and producing wood. Wood is Western man's largest crop. This attitude is much less prevalent in the East, even in much of the Near East where forests used to grow. Yet it was in classical Greece that the first scientific botanist, Theophrastus of Eresus, classified the land vegetation known to him into just four categories: trees, shrubs, undershrubs, and herbs. Agreement on what to call a tree, or how many trees are the minimum to make a forest, is more general than concepts relating to Theophrastus' other categories of plants.

A full analysis of life forms in plants came only in 1934, when a Danish botanist, C. Raunkaier, produced a stimulating book on the subject. He sought to provide a comparative base that, like the word tree, would be independent of specific identification, yet would relate the size, shape, structure, and permanence of the plant body in terms of adaptations important during the seasons of the year. Dissatisfied with adjectives such as annual, biennial, perennial, woody, non-woody, terrestrial, and aquatic, he focused on the site of the potentially immortal growth centers and separated out ten different categories:

A. *Unicellular, with no seeds or buds*

    I. Phytoplankton: plants microscopic or nearly so, suspended in water;

    II. Phytoedaphon: plants microscopic or small, within the soil;

    III. Endophytes: plants microscopic, enclosed within other organisms, such as the bacteria in root nodules and the algae in coral polyps;

B. *Multicellular, with growth centers in seeds or buds*

    IV. Therophytes: annuals, bridging adverse seasons by means of growth centers located only in the seeds;

    V. Hydrophytes: biennials or perennials with growth centers in buds below the surface of water;

    VI. Geophytes: biennials or perennials with buds below the surface of the earth, such as many ferns, daffodils, asparagus, and rhubarb;

    VII. Hemicryptophytes: biennials or perennials with buds at ground level, such as iris, English ivy, onions, and most grasses;

    VIII. Chamaephytes: low shrubby perennials with buds no more than 10 inches above ground level;

    IX. Phanerophytes: perennial shrubs and trees with buds more than 10 inches above the ground, subdivided according to ultimate height into micro-, meso-, and megaphanerophytes; forest trees are megaphanerophytes;

    X. Epiphytes: plants with their point of attachment to the substratum as well as all buds above ground, generally atop a phanerophyte.

Armed with this logical classification, Raunkaier hoped to establish "statistical plant geography" by comparing the various terrestrial biomes according to the relative proportions of plants in each of his ten categories. He referred to this measure as the "spectrum of life forms." But although his analysis was so much more definite than earlier qualitative estimates, it did not help identify the features that determine which areas of the world are dominated by forest trees, rather than by thornscrub, grasses, or desert plants.

To produce a forest, several conditions must be fulfilled. The climate must be moist at all seasons, never severely dry. The subsoil must contain a permanent reserve of available water. The maximum rainfall must come during the growing season, not during a period of relative dormancy, such as winter. Fires must be infrequent, and large herbivorous animals too few to interfere seriously with the survival of young trees.

Once established, a forest exerts a strong influence toward perpetuating its own ecosystem. The trees reduce the amount of light reaching ground level, shield organisms from the effects of wind, and conserve substantial amounts of moisture in and above the soil. The relative humidity below the interlacing leafy branches is much higher than above them, and the temperature varies much less through each day and each year than in places where the land is open to the sky. Less obvious is the higher concentration of carbon dioxide in the air below the forest canopy, due to the activities of animals and especially the decomposers. For photosynthesis this feedback is immensely important in stimulating forest growth.

The woody vegetation to which Raunkaier applied the name phanerophytes has evolved through many changes since it appeared in the late Devonian as the earliest known life form of plants on land. The swamp forests of the Carboniferous must have been extraordinarily productive to have left so much fossil fuel as organic matter that escaped complete decomposition. Yet among the associated animals, insects may have been the chief beneficiaries for millions of years. Their mobility, resistance to desiccation in exposed situations, and high biotic potential make them prime converters of woody plants into materials a vertebrate animal might use as food. Except for insects, insectivorous vertebrates, and the predators and parasites that benefit from these, few animals inhabit any mature forest. As soon as the rate of photosynthesis is matched by the rates of consumption and decomposition, there is nothing left over as a yield that can be removed without destroying the forest resource.

In the modern world, more than a fourth of the land area is still forested—a total of about 15.2 million square miles. Of these forest areas, about 8.2 million square miles are regarded as harvestable. At least as much former forest has been felled and converted to pasturelands or to other uses. Yet these changes are not easy to maintain. In many parts of Scotland, for example, the trees that are the normal climax vegetation have been cut and the growth of grass encouraged as food for sheep. The sheep eliminate the seedlings of trees that might colonize, but generally keep the grass too short to compete with the less palatable heather and gorse. As these woody shrubs become the dominant vegetation, far less food is available for the animals upon which so much of the Scottish economy depends. Yet the grass does not recover if the sheep are removed for a few years. Instead, the seedlings of trees appear, shade out the heather and gorse, and reinstate the forest. Only with less grazing and more frequent fires than Scotsmen now

Where the range in temperature is small and water and sunlight are abundant, vegetation can develop into a rain forest. Competition for water, light, and mineral nutrients reaches a high level. Far more species live in tropical rain forests than in temperate ones. Along the Oyapogue River of French Guiana (left), heavy bromeliads perch on every available surface. Vines send down long roots toward the scanty nutrients in the soil. But most of the useful nutrients are held within the tissues of the vegetation. Prevailing winds from over the relatively warm waters of the eastern Pacific maintain high humidity and drop more than 100 inches of rain upon Olympic National Park (far left) each year. Along the Hoh River, the temperate rain forest supports woody vegetation and epiphytes from algae and lichens to flowering plants (bottom, left).

permit can a grass cover be maintained on forest land.

The Tropical Rain Forests

More kinds of trees grow close together in the tropical rain forest than anywhere else on earth. Forests of this type cover a major portion of tropical America, extending over the vast basin of the Amazon and its interconnections with the Orinoco River, southward to the boundary between Brazil and Argentina, northward through lowland Central America to Mexico's state of Chiapas. Smaller areas occur under similar climatic conditions in the basin of the Congo, and in Asia with a center in Malaya but outliers as far west as the vicinity of Bombay, and east through northern Australia to some of the islands in the South Pacific.

Many areas of rain forest on lowland grade on mountain slopes into a somewhat different montane rain forest, and finally an elfin woodland. If the peak is high enough, this may be followed by an alpine tundra, and finally the snowfields and glaciers that prove conclusively how much cooler the climate is at high altitudes. Far up into the montane forest, the rainfall may continue to exceed 80 inches annually, well distributed throughout the year. Yet the prevailing winds that bring the moisture differ in temperature, making the lower montane forest a subtropical one, the upper montane a temperate rain forest, and the elfin woodland a strange place of chilly mists often described as a "cloud jungle."

Along the rivers in their upper reaches, the trees grow dense and tall because their roots have plenty of moisture. Only a few hundred yards from the river bank on each side may be a grassland if the rainfall is inadequate to support forest growth. The strip of forest is called a "gallery forest," and represents an extension into higher, drier territory of the lowland, wetland type of vegetation.

Along the coasts, the tropical rain forest is replaced by dense stands of mangroves. These tolerate salt water and the rhythmic changes of water level daily due to tides. The adventitious roots of mangroves trap an immense amount of debris and help extend the land as well as protect it from erosion by storms. Many of the animals that frequent the edges of rivers, where the water is fresh, range also among the mangroves in search of food and shelter. But toward the sea the variety of life diminishes greatly, just as it does up the mountainsides.

The richness of the forest flora in the tropics can be visualized by comparing the trees on 1¼ acres of undisturbed rain forest in northeastern Australia with those on a similar acreage of beech–maple forest in southern Michigan. When only trees exceeding an inch in diameter were counted, the Australian forest was found to have 1,261 individuals representing 141 species, the commonest of which accounted for only about 9 per cent of the total basal area. The Michigan forest had about the same number of trees but only 10 species, and among these one accounted for 51 per cent of the total basal area in the stand.

In the tropical rain forest, groves of a single species are rare. There are no mahogany forests or teak forests, for no dominance develops. Usually each tree is remote from another of its kind and age. Solitary giants may tower 75 to 100 feet above their mature neighbors, attaining a height of 300 feet or more. Below these ''emergents,'' the forest canopy is extremely uneven, although it is laced together by clambering, flowering vines. This is the stratum that is most frequented by monkeys and other climbing mammals, by snakes and lizards, frogs and toads, by insects in tremendous variety, by bats (especially the fruit-eating kinds) at night and by birds in daylight hours. So rich a fauna finds food, living spaces, and concealment on massive tree branches that grow almost horizontally. Each branch is loaded with literally tons of epiphytic plants. The epiphytes include ferns, clambering clubmosses, true mosses, liverworts, lichens, spectacular orchids and, in the Americas, various water-catching members of the pineapple family known simply as bromeliads. Virtually all of these plants and most of the animals have no stage of development that is closer to the ground.

An understory of extraordinarily slender trees generally rises with scarcely a side branch to a height of 50 feet or more. Mostly these are juveniles of the canopy trees and the rare emigrants. Few are strong enough to tolerate a wind, or are ever exposed to any gust. Some are barely thicker than the climbing vines, called lianas, and the adventitious roots that swing and sway, extending vertically 100 feet or more, carrying dissolved nutrients from deep in the forest floor to foliage at the level of the high, outstretched tree limbs.

Often the soil stratum in the rain forest is bare mud, with few low-growing plants affording food to animals that cannot climb. The principal forage for the wild pigs, such as the peccaries of the American tropics, and other denizens of the forest floor consists of great limbs that break off and fall with their burden of epiphytes and associated small animals. Prompt inspection of the broken limb invariably reveals that it has been weakened disastrously by boring insects and fungi. Within a few days only the woody parts remain, and these too disappear in a month or so as termites and decomposers complete their reduction of the organic matter. Often a decomposing limb of large size or the remains of a fallen tree trunk becomes a seed bed for seedlings of its own species, giving them a chance to produce roots deep into the lateritic soil where they may reach nutrients.

Sprouting from a fig seed dropped by a bird, an epiphytic fig (Ficus) produces its own foliage and clasping roots. Its woody roots then grow downward, holding to the bark of the tree, and eventually reach the soil. Gradually, the roots thicken and join together, forming a sheath of roots around the tree, while the foliage competes for light in the forest canopy. When the tree dies, the "strangler fig" inherits its place in the rain forest.

Each tree species seems to synchronize its flowering and fruiting, thus gaining fullest use of the animals that serve in pollination and help in seed dispersal. These include many mammals and birds as well as insects. Many of the flowers are highly adapted in ways that limit pollination to one or just a few related species of animals. This reduces wastage of pollen and improves the chance of cross-pollination. Since most of the plants are long-lived woody kinds that flower year after year, the chance of eventual success in fruiting and germination of seeds is good despite the considerable distance between one plant of a species and the next. As a corollary, the animals that visit flowers and help in seed dispersal have a succession of different species to nourish them throughout the year. This increases opportunities in a way of life that has no true counterpart in other biomes.

Without genuine challenge from the physical environment at any season, the plants and animals in the tropical rain forest have evolved a wonderful multiplicity of niches. These divide the biome into horizontal strata, into annual periods of a week or less for conspicuous activity, and into more social or symbiotic relationships than are to be found elsewhere in the world. Competition within a species is virtually absent, except among the pioneers that spring up in an abandoned clearing or where a giant tree has fallen and smashed a great hole in the forest, letting sunlight reach the ground. Along the broader rivers, the canopy and the tangle of vegetation comes within reach. These are the place to which the word "jungle" properly applies. The great interior of the rain forest is almost empty, with all of the activity high overhead.

The American moose eats the vegetation from the bottom of beaver ponds. During the winter, the moose uses its long legs and spreading feet to wade through the snow and reach nutritious buds and young twigs. The animal is a characteristic member of the fauna in the taiga, which Americans often refer to as the "spruce-moose" community or biome.

The subtropical rain forests of the lower montane elevations may still be rich in variety, but rise only to a canopy 70 to 100 feet above the ground. On one 100-acre tract of this type in Trinidad, about 16,000 trees exceeded 1 foot in girth, 800 over 6 feet, and 80 over 10 feet. Virtually all of them bore simple, elliptical, evergreen leaves, somewhat leathery and shiny, ending in a downturned point called a "drip tip."

The temperate montane rain forest in the tropics grows to only about 60 feet in height, and characteristically shows considerable wind damage. Almost as much water comes in the form of mist as in rain. Epiphytes have their greatest development here, and the ground bears a dense herbaceous vegetation composed of relatively few species.

In elfin woodland, the flora is still more limited, the trees gnarled and twisted, pruned by the wind at a height of only 15 to 25 feet, but festooned with lichens and mosses. Often an understory of tree ferns and small palms forms a closed canopy 6 to 10 feet above the soil. The elfin woodland may end abruptly just over the crest of the mountain because, in the rain shadow of the prevailing winds, adjacent slopes receive so little moisture that they are almost free of vegetation.

The Temperate Forests

Beyond the tropics, the cyclic variation in the length of day and in the angle at which sunlight reaches the surface of the earth brings seasonal changes that have no counterpart closer to the Equator. At 40 degrees of latitude, which is the approximate position of Salt Lake City, Denver, Philadelphia, Madrid, Istanbul, Peking, Valdivia (Chile), and the Bass Strait between

246

Australia and Tasmania, the number of hours between sunrise and sunset varies from 9.3 at winter solstice to 15.5 six months later. Temperatures, whether maximum, average, or minimum, vary greatly between summer and winter. This influences the amount of moisture that wind carries away, and hence the residue from precipitation that is available to the roots of trees. Exposure to sun differs too in temperate regions, with greater warmth by day and a longer growing season each year on land that slopes downward in the direction of the Equator than on land whose downslope is toward the nearer of the earth's poles.

The Coastal Coniferous Forests of Western America

The tallest trees in the world, and some of the runners-up as well, grow near the Pacific coast between central California and Alaska. Those at the southern end of this range are predominantly the coastal redwoods, few of which perpetuate themselves more than 30 miles from the ocean and hence beyond the reach of dense coastal fogs. Farther north, where the rainfall increases to as much as 120 inches annually and lower average temperatures reduce the evaporative losses, the forests are broader and more varied. In the Puget Sound region, the dominant evergreens are western hemlock, western arborvitae, grand fir, and douglas-fir. Farther north the Sitka spruce becomes predominant. Yet the spacing of the mature giants generally allows sunlight to penetrate, giving energy to an understory of shrubs and small trees. The ground is covered by mosses, ferns, and other vegetation that needs high humidity. But the soil is unsuitable for a spring ground flora.

Often the coastal coniferous forests are referred to as "temperate rain forests." The one conserved in Olympic National Park does have a total annual rainfall comparable to that in many tropical rain forests. But slower decompostion lets the soil become a podzol rather than a latosol. It supports none of the spectacular wealth of species, either plant or animal, characteristic of the tropics. Almost solid stands of trees of a single kind are common. Competition within a species is often severe, as a social relationship rarely encountered near the Equator. The number of species in the temperate forest is a good indication of the number of habitats available, and of the opportunities for new species to evolve or move in. It reveals also a great vulnerability of the ecosystem. Removal of a key species can severely change the whole community.

Today the spectacular forests of coastal redwoods are doubly vulnerable. Those that have been saved from the saw are ceasing to reproduce themselves because fire control is so efficient. Without a brush fire once or twice a century to clear out the undergrowth, while merely scorching the thick bark of the redwoods, the vegetation between the giant trees grows so dense that redwood seedlings cannot get a start. Yet where the big trees have been felled and dragged away, grasses spread. Young redwoods, tree seedlings, and many other kinds of life in this forest community perish from drought. The explanation lies in the ability of the evergreen needles on tall redwoods to condense moisture from the coastal fogs and to drip it to the forest floor. As much as 50 inches of water annually have been recorded in a redwood forest where rainfall above the trees was in the 10- to 20-inch range. Loss of an occasional tree from old age in undisturbed forest has no such consequence because the opening in the forest is small. Soon it fills with pioneers, then seedlings of the climax species.

The Temperate Deciduous Forests

In parts of the world where winter brings cold weather for several months, trees with broad leaves are challenged by a drastic reduction in the availability of water. Even if the moisture remains unfrozen, it becomes viscous. It moves slowly in the soil and in the conducting tubes of plants. The cold also slows the growth of roots toward more moisture in the soil. The broad-leaved trees in these regions have evolved the habit of shedding their foliage almost synchronously, then standing bare all winter. This reduces greatly the area of surface through which moisture can be lost to wind, and to which freezing rain (or heavy, sticky snow) can cling, causing mechanical damage to the tree. The adaptation is effective only where the growing season is three months long or more, allowing time for new leaves to expand, attend to photosynthesis for the year, and then drop off. The temperate deciduous trees have a thicker bark than most in the tropics, and waterproof scales over their winter buds as additional features that reduce the danger of winter desiccation.

Forests of this kind were once widespread in China and Japan, in Europe from northern Spain to southern Scandinavia and central Russia, and in North America from Minnesota to Nova Scotia and southwestward to Missouri. The greatest development was in the southern Appalachian Mountains, particularly the Smoky Mountains of Kentucky and Tennessee. There the forest in many places still presents a closed canopy of interlocking leafy

Adaptive features make advantageous many relationships between plants of unlike species. The epiphytic wild pineapple (Tillandsia fasciculata) *of the American tropics maintains a roothold (upper left) in the rough bark of a tall tree. It shades none of the foliage of the supporting species and carries on its own photosynthesis. The channelled lichen* (Ramalina calicaris), *another epiphyte (upper center), often forms a dense coating over the bark of the Swiss mountain pine* (Pinus mugo). *The Indian pipe* (Monotropa uniflora) *of American woodlands (upper right) appears to derive its nourishment from soil fungi that are mutualistic mycorhiza around roots of hemlock and other trees. Many of the saprophytic fungi, whether gill-bearing (lower left) or releasing their spores through minute pores, gain nutrients from dead wood as decomposers, but in early stages serve in mutualistic partnerships with living roots. The fungus* Penicillium *(lower right), which decomposes nonliving organic matter of plant origin, became famous as an example of amensalism because it secretes antibiotic compounds that reduce competition from bacteria for its food resources.*

*Adaptations promoting the beneficial
transfer of energy and inorganic
nutrients show almost endless variety.
With minute jaws at the tip of its long
proboscis, the nut weevil* (Balaninus)
*opens a hole through which to lay an
egg inside an acorn, hickory nut, or
other hard-shelled fruit (top, left); the
larva feeds on the nut. The eastern
garter snake (bottom, left) and other
snakes slide through low vegetation and*

branches 80 to 100 feet above the ground, where about 25 species of trees flourish in a mixed stand. Beech, hickories, oaks, basswood, and red maple predominate, with occasional tuliptrees attaining a height of 150 feet. Below these is a lower canopy of trees with genetic limitations for lesser tallness: flowering dogwood, redbud, holly, and hornbeam, seldom rising higher than 40 feet. A shrubby stratum at half this height consists of benzoin, pawpaw and witch hazel. Below this, growing 6 to 24 inches tall, is a distinctive stratum of spring wildflowers, including bloodroot, trillium, and violets. They bloom soon after the snow melts, and within four to five weeks are ripening their fruits—about the time the leaves on the trees above expand and close the canopy for the season, shutting off most of the light at ground level.

The spring ground flora is a unique feature of the temperate deciduous forest. It temporarily overspreads a thin mat of creeping evergreen plants, including mosses, clubmosses, and low-growing members of the heath family. Herbaceous fronds of delicate ferns often take the place of the spring ground flora by the time the canopy overhead has blocked out most of the sunlight.

Slopes with an exposure in the equatorial direction and river bottoms tend to demonstrate their longer growing season by supporting plants characteristic of lower latitudes. Slopes with a poleward exposure and higher elevations have a flora like that of higher latitudes. Hence a map that shows the contours is essential for recording the variations in the forest flora and in the animals that obtain their food from the vegetation directly or indirectly.

Stratification of the animals is generally

less impressive in a temperate deciduous forest than in a tropical rain forest, because a good many kinds can be found at each level. But few species range far vertically. Predatory mammals climb to considerable heights in search of prey, but their choice of den sites for young is more limited—usually near or at ground level. The majority of birds, both in number of species and of individuals, frequent the shrubby layer where insects and fruits are more varied. Europeans have recorded that in their deciduous forests 29 per cent of the birds nest in the canopy, 25 per cent in the shrub and herb stratum, 15 per cent on the ground, and the remaining 31 per cent in special situations such as tree holes, which vary greatly in height. North American birds show similar preferences for particular strata. Moles and shrews remain close to or in the soil, as do the white-footed forest mice.

Before man began setting fires and felling trees in North America, the deciduous forest extended from the Atlantic coast to the drier fringes of the Mississippi River basin, and maintained to the west a shifting boundary against grasslands. In wet periods it advanced westward and in dry ones retreated eastward. To the north it became progressively more intermixed with evergreen coniferous trees until the term deciduous no longer applied. The biome, instead, became that of the boreal coniferous forest, the taiga.

The Taiga

From Norway to eastern Siberia across Eurasia, and southward as far as northern Poland, a great forest of spruce, fir, and larch forms the biome to which the Russians give the general name taiga (pronounced tie-gah). A counterpart, with

closely related species of the same genera, continues in the New World from Alaska to New England and northern Newfoundland, and down the mountain chains for some distance at increasing elevation. No other part of the world that life inhabits receives so much snow. The larches escape a confrontation with this winter hazard by shedding their needles. The spruces and firs benefit from their pyramidal shape and decurrent branches, which let most of the snow slip off into the low places between the trees. Low branches rest on the ground and support upper ones until wind and sun free them from their burden.

The coldest temperature recorded in any place that man inhabits is from this biome: −90°F (−68°C) in 1892 at Verkhoyansk, U.S.S.R., at 127° east longitude just north of the Arctic Circle. Yet the summers are long and warm enough to thaw the soil completely, raising the average annual temperature well above the freezing point. The thick accumulation of resinous needles, which forms a duff below the coniferous trees, acts as insulation for the podzol produced by slow decomposition. The soil remains cool all summer, and the roots of the conifers seldom penetrate far below the surface. Instead, they tend to interweave and produce a shallow layer of roots, each tree being held down by its neighbors on all sides.

The acid soil is generally soggy, favoring growth of peat moss, other mosses, clubmosses, and low-growing members of the heath family such as blueberries, cranberries, wintergreen, and Labrador tea. Where bog water drains into streams and stains them a dark coffee color, herbaceous plants are more numerous. These nourish in summer both the moose for which the taiga is famous and the snowshoe hares in

The taiga, or northern coniferous forest, of Siberia extends for thousands of miles across Eurasia. It is composed mostly of firs, spruces, pines, and larches, with broad-leaved, deciduous birches and aspens along the margins of the many rivers. Although summers are warm enough to thaw the soil completely, the growing season is too short, and the winter too severe and snowy for most crop plants or domestic animals. Consequently, the human population is generally small and composed mostly of people who crop pulpwood or who engage in mining operations.

255

their dark pelage. In winter, the moose and the hares dig down through the snow to reach the berries and evergreen foliage of the wintergreen and Labrador tea, or browse on the conifers. Squirrels, crossbills, and siskins find seeds in the cones of the conifers, and a refuge from storms under the drooping branches.

The short summers and long winters in the taiga do not limit productivity as much as might be expected. Soil moisture is plentiful from melted snow and summer rains, and gains protection by the almost perfect windbreak provided by the dense evergreens. The green needles on spruce and fir retain their moisture well, and carry on photosynthesis earlier and later in the year than any broad-leaved tree could manage. The number of species is small but the annual increment is large. Yet as in the tropical rain forest, most of the mineral nutrients are bound up within the standing trees. If these are harvested and hauled away, only deliberate replacement of the nutrients seems likely to sustain the yield in this biome. The taiga is unsuited for preferred crops that yield food and fiber because these require a longer growing season and a neutral or alkaline soil. Remoteness of market centers is an economic factor operating in the same direction.

The trees of the taiga prevent wind from having much effect at ground level, but their pointed tops interfere little with air movement. Conifers produce most of their cones high up and rely upon wind pollination. This habit, which is a tradition among gymnosperms dating back before the evolution of flowers or of pollinating insects, birds, and mammals, can be followed efficiently in the boreal forests (although not in the tropics) because woody plants of the taiga show so little stratification and grow in almost solid stands.

At the highest elevations to which the taiga extends on mountain slopes and along the northern boundary, the trees are dwarfed and contorted. Circumpolar winds prune away new growth that extends above the hard-packed snow, and pelt the branches with ice, killing buds on the windward side. The boundary is known as tree line, or as timber line although the timber there is virtually useless. The twisted trees are often referred to as krummholz, from the German term ("crooked wood") applied to them high in the Bavarian Alps.

The present location of the taiga is relatively new, perhaps no more than 10,000 years old. Before that time, it was farther south, still as a transcontinental band bounded on the north by tundra and on the south by deciduous forest or grassland. The two early and two late expansions of the great glaciers during the Ice Age forced the taiga to shift southward. As the ice retreated, the conifers colonized the bared rock and glacial debris.

In all of this time, summer drought has rarely been a limiting factor. Instead, the boreal forest resembles the temperate deciduous forest in summer and the tropical rain forest at any time in one particular; during its growing season it releases by evapotranspiration almost as much moisture into the air as would be liberated from the surface of open water. This rate is possible—and so is tree growth in any pattern suggesting luxuriance—only where the natural precipitation is at least equal to the potential loss of moisture into the passing winds.

16 Grassland communities

Thinking about the plants that constituted the greatest natural wealth in his state a century ago, Senator John James Ingalls wrote in the *Kansas Magazine*, "Grass feeds the ox; the ox nourishes man; man dies and goes to grass again; and so the tide of life, with everlasting repetition, in continuous circles, moves endlessly on and upward, and in more senses than one, all flesh is grass."

If all members of the grass family (Gramineae) were embraced in Ingalls' claim, they would include not only the forage for most of the world's livestock but also the grains that are so important as human food—rice, wheat, millet, oats, rye, and barley—and for good measure maize, sorghum, sugar cane, and bamboo shoots.

So important have members of the grass family become to man in the last 10,000 years that he has deliberately expanded their geographic scope and tailored their natural communities to suit his plans. To make space, this has entailed destroying large tracts of forest at a rate even faster than other areas were paved and built upon. Consequently, more than a fifth of the total land area on the earth now supports grasses. No other single family of plants gives character to so much territory, from the desert edges to alpine meadows, from marshes to forest margins.

Over a substantial part of this total area, the members of the grass family spread without human help or hindrance. They colonized and competed for the semiarid portions of the world, beginning about 60 million years ago. Grasses developed along with nongrass vegetation, grazers and seed-eaters, predators, scavengers, parasites and decomposers. Together they built a productive ecosystem where almost nothing had lived before.

In meeting the diverse ecological limitations in so many parts of the world, the grasses evolved into more species than are found in any other family of plants except Compositae (the daisy family) and Orchidaceae (orchids). Yet a few original types contributed disproportionately to the success of grasses in providing the earth with an additional biome. Nearly half of the known kinds of grasses belong in just a dozen genera, each with 100 or more species whose differences correspond to distinctive niches in the various habitats they occupy.

Most grasses are easy to distinguish from other plants, and their adaptive features are surprisingly uniform. Their fibrous roots, their untapering, jointed stems, and their narrow linear leaves follow a single pattern. Persistent growth centers provide for elongation of the stem at every joint, and for continuous extension of the leaf blade with its parallel veins. The loss of a stem tip interferes minimally with enlargement of a grass plant, whereas a nongrass is slowed seriously while scar tissue forms and then a new terminal bud develops, from which new stem and leaves can arise. If a herbivore (or a lawn mower) clips off much of a grass blade, more grows out with no need to open a new leaf bud, as any other plant would have to do in replacing its photosynthetic organs. The buds of grasses are at or below the surface of the ground, generally out of danger from grazing animals, excessive sun, severe desiccation, and wild fire. Customarily the leaves rise at a steep angle, intercepting light for photosynthesis early and late in the day, but spreading the sunshine at noon (and its heating effect) over a large area, thereby reducing the stress of a hot summer day. Until the hour when the reproductive parts

Giraffes seem out of place on the grass-lands of Kenya without the flat-topped acadia trees on which they feed so easily. The thorn forests that once grew here were converted into a savanna and then a plain by the activities of many elephants. When the trees have been killed, the elephants depart. Meanwhile, the giraffes survive on lower vegetation and benefit from a clearer view of approaching predators.

are ready for use, they are protected in a grass flower by a distinctive sheath (glume). It opens on schedule to expose the stamens while they dust pollen into the wind and the pistils catch the pollinating dust.

In any grassland that remains lightly grazed, infrequently burned, and totally unmowed and unplowed, a layer of dry debris known as "fresh mulch" accumulates atop the soil. This material includes dead blades and upright stems of grasses, empty grain heads, and various dead parts of the nongrass plants that people on the western grasslands of America refer to collectively as forbs. As much as 5,000 pounds per acre may build up. Yet, until the mulch fragments mechanically into small pieces and comes into actual contact with the mineral matter of the soil, it holds little moisture and shows few indications of decay. It does protect the soil from the force of rain and wind, and insulates the earth to some extent. A year or two may pass before the fragments of mulch become broken and compacted, letting them retain some water and be acted upon by decomposers. Ordinarily the material takes about four years to progress through successive stages as "humic substances" in the soil before it disappears in the great biogeochemical cycles. This natural cycle may maintain 9,000 pounds of organic material per acre in the A and B horizons of grassland soil.

The weight of fresh mulch decreases in a series of dry years, and then so do the humic materials. The grasses and forbs change too, those perennials with shallow roots being replaced by deep-rooted kinds and annuals. These natural adjustments are reversible. They are part of the adaptive repertoire of grassland plants and native animals, which have had a long history of

258

South of the great deserts in Africa, prey animals, particularly antelopes, are still sufficiently numerous on the grasslands and thornscrub to support a healthy population of lions (left). These big predators have little effect on the African elephants and black rhinoceroses (right) that forage selectively on the vegetation. Predatory man and his herbivorous, water-demanding livestock are the principal competitors of the large native mammals in this area.

surviving even the slow cycles in climatic variation.

The forbs play an important, although scarcely appreciated, role in the flora of grasslands. The legumes, such as vetch and lupine, house in the nodules on their roots the *Rhizobium* bacteria that fix atmospheric nitrogen into the form of nitrogenous nutrients; these dissolve and diffuse through adjacent soil where grasses and other plants can benefit. The composites, such as asters and goldenrods, and members of other families, such as pigweed (*Amaranthus retroflexus*) and tumbleweed (*A. graecizans*), are generally regarded as worthless because they have a less direct effect. Yet their roots and products of decomposition support the nutrition of soil organisms, whose activities do benefit the grasses. The forbs, like the grasses, are limited to a very few strata: ground-level wild strawberry, violets, and dandelion;

wild mustard and Eurasian daisy barely higher than the short grasses; asters and goldenrods conspicuous in the late summer when the tall grasses attain their full size.

Grassland animals are equally limited in variety. The most numerous invertebrates are insects, chiefly herbivores. They include leafhoppers and spittle insects, aphids (plant lice), and grasshoppers that suck juices or chew foliage, and then avoid the dry season and the cold as dormant eggs. Ants, such as the agricultural ant, store food underground and reduce their activity in adverse weather. Spiders, with few shrubs in which to stretch webs, tend to be hunters rather than web-makers and to find refuges underground during winter and droughts.

The birds of grasslands find so few perches higher than a grass that they seem to have become adapted either to making their defense of territory an aerial singing bout (as is customary among larks and bobolinks) or a seasonal social display in which the strutting, competing males attract a harem of ready females (as is well known among the distinctive prairie chickens of North America). During the growing season, the birds become largely insectivorous or supplement an insect diet with soft parts from the prairie forbs. For the dry season and winter, the birds must either migrate or subsist on what seeds they can find. On cropland, the carrying capacity in off seasons is generally so reduced that prairie chickens, quail, and other nonmigratory birds tend to disappear.

The mammals of the grasslands are mostly burrowing rodents, which often raid the grain stores of ants in winter, and cud-chewing ungulates in whose four-part stomachs live cellulose-splitting microbes that can get nourishment from dead plant material. Horses, whose cellulose-digesting partners live in the large intestine, evolved in this same biome. Like the pronghorns and true antelopes, they became long-legged and fast-moving, sharp-eyed and able to run from predators that could be seen approaching from a distance.

Replacement of the native life with domesticated grazers and field crops has been so extensive that the grasslands of the world have been altered more by human efforts than any other ecosystem. Too often the economic gains from raising meat and grains have led to practices that could not be sustained during dry phases of the weather cycle—a variation that brings drought about thrice each century.

Among the best-known debacles from excessive use of grasslands had its beginnings prior to the prolonged drought of the early 1930's when much of the driest part of the Great Plains (particularly in Oklahoma) was plowed and planted to annual crops. Other areas were heavily overgrazed

Domestic herds of alpacas on alpine grasslands in Peru provide many raw materials as well as meat to the indigenous people. Like the domestic llamas and the wild guanacos and vicuñas, the South American camelids whose forebears came from North America, arrived at about the time other camelids crossed the Bering land bridge into Asia.

by livestock. Both agricultural operations broke up the sod, prevented grasses from maintaining a firm mass of interlocking roots, and deprived the soil of most of its nourishing mulch. When the annual precipitation diminished and winds raced over the bared soil, the thin scattering of mulch blew away and the topsoil followed, leaving dunes in some places and exposed subsoil in others. After 40 years, much of the famous "Dust Bowl" has not yet fully recovered despite millions of dollars spent in attempt to reclaim the land. Elsewhere, the original plant communities tended to restore themselves as soon as rainfall returned to average rates.

The Open Lands With Many Names

The soft rustling of stems and blades and grain heads bending with the wind, the songs of hovering birds, the general solitude of broad grasslands extending to the horizon in all directions, combine in a romantic picture that applies equally in many parts of the world: the rich grain fields of the Ukraine, the desolate steppes of central Asia, the vast savannas of equatorial Africa (which extend almost coast to coast), the desolate karroo of South Africa, the semiarid agricultural areas encircling the great deserts of central Australia, and the immense grassclad tracts in South America known as campos, chacos, pampas, punas, or understandably as altiplanos ("high plains"). Each has its distinctive characteristics, native flora and fauna, and history of human exploitation. Each resembles in many ways the great Central Plains of North America, from midway in Alberta and Saskatchewan southward through the Dakotas, Nebraska, and Kansas to eastern Texas, spreading along the Gulf Coast into Louisiana and south into Mexico.

Largest by far is the grassland that extends across Eurasia for 2,500 miles from the Ukraine to the Altai plateau in southwestern Mongolia. Few fragments of this vast realm are still clad in the native feathergrasses that grew tall when, less than 3,000 years ago, Eurasia still had great herds of wild horses, bison, saiga antelopes, and other animals that now are extinct or vanishing. An immense counterpart of the North American prairie, it was equally the home of burrowing rodents and migratory locusts. But interspersed among the grasses of the Old World were red tulips, magenta peonies, blue sage, and many other plants that have been developed as horticultural delights. Today its wildlife is mostly the smaller denizens, such as the ground squirrels (susliks) on the clay steppes, the jumping jerboas and burrowing hamsters of the sand steppes, the sparrows and larks, and the foxes and Pallas cats that stalk the young, the sick, the old and unwary.

Across Africa both north and south of the great deserts, grasslands that were widespread and productive until Roman times have largely disappeared. The grasses were destroyed by the goats, sheep, donkeys, and camels of nomadic people. Scattered annual grasses, such as the triple-awns, have replaced the perennials—even the sandburs (*Cenchrus*) except where rainfall is somewhat higher toward the Equator. On the highlands from Mauretania in the west to the Nile Valley in the Sudan, the semiarid land is a vast steppe where domesticated herds and frequent fires (many of them set by men) prevent any succession toward a natural climax. Yet in many places, the elephant grass (*Pennisetum purpureum*), which is a relative of millet, attains a height of 10 to 12 feet. In

others, lower grasses such as red oat grass (*Themeda*) cover the land, supplying the principal food of mixed herds of gnus (wildebeests) and zebras, antelopes, and gazelles of many kinds, and small flocks of ostriches. Troops of baboons search for insects and fruit. Cheetahs pick out and run down individual prey. Prides of lions cooperate in ambushing their victims. But the prominent plants are often spiny acacia trees, each shaped like an inverted cone. The trees are scattered, making the land a savanna. Giraffes browse from the upper branches, while smaller animals find shade and some concealment below.

On many of Australia's semiarid lands, particularly in the south, nothing taller than a short grass is evident. These are the nullarbor ("no tree") plains. Elsewhere the annual triple-awn grasses are mixed with perennial mulga grass (*Danthonia racemosa*, of a genus found widespread also in South Africa) and studded with harsh acacia shrubs called mulga (*Acacia*

aneura); this is mulga scrubland. It is home to red kangaroos and emus as grazers, marsupial wombats as burrowers, and small flocks of cockatoos as fruit-eaters. In combination with dogtooth grass (*Cynodon*) and one called *Spinifex* from the sharp, elastic spines on its grains, some of these same grasses grow where certain low shrubby gum trees called mallee (*Eucalyptus*) make it a mallee scrubland. Each of these scrublands and grasslands grades into outright desert toward central Australia.

South America's grasslands are more widely separated, each sustained by a slightly different version of semiarid climate. The famous pampas of Argentina, which have outliers on the Falkland Islands, are well supplied with panic grasses and tall tufts of the valuable pasture grasses in genus *Paspalum*. The still taller pampas grass (*Cortaderia*), which is often planted elsewhere in the world because its handsome plumes of flowers and seeds rise 8 to 12 feet above the ground, grows chiefly along drainage channels. Formerly this was the home of the furbearing nutrias (or coypus) along the water courses, of pampas deer, guanacos, and the strange South American rodents known as plains vizcachas. Rheas roamed in flocks comparable to the emus of Australia and the ostriches of Africa. During the winter of the Northern Hemisphere, great numbers of songbirds and shore birds arrived after long migrations. Today the pampas are cattle country, and much of the wildlife is gone.

Toward Patagonia the forage grasses include many kinds of *Agrostis*, with tussocks of *Poa* and bushes rarely as much as three feet tall. To the west, these grasslands dwindle into the desert that occupies the rain shadow of the southernmost Andes.

Farther north, on the mountain slopes, a slightly greater precipitation maintains the sparse grasses of the windswept punas. These are chiefly fescues (*Festuca*) and bentgrasses, which offer little shelter and only a small amount of grain to birds. Yet along the ridges, the fleet camels known as vicuñas graze with a minimum of competition. Mountain vizcachas, chinchillas, and guinea pigs are native. These animals rarely venture into the shallow, grass-clad valley between the eastern and western divisions of the Andes chain in Peru and Bolivia—the altiplano where the prominent bunchgrass (*Stipa ichu*) is used by the Indians for forage and thatching.

The prairies of Uruguay, the chacos of Paraguay, and the Brazilian campos all have their distinctive native grasses, many of which have been crowded aside to make space for introduced kinds from the same and other continents as food for domesticated livestock, or as cereal grains that man can use directly.

The ecologist becomes impressed by the convergence in habits related to the grassland habitats, as shown by unrelated kinds of warm-blooded animals on different continents. The burrowing vizcachas and other rodents of South America have their counterparts in the marsupial wombat of Australia, the marmots of Eurasia and North America, and the ground squirrels of these two land masses and of Africa too; these animals feed in the open, mostly at night. The leaping herbivores of Australia are the kangaroos and wallabies, of Africa the strange springhaas, of Asia the jerboas, of South America the vizcachas, and of North America the various jack rabbits. Each region has its burrowers that feed underground: the marsupial mole in Australia, the golden moles of Africa, the tuco tuco of South America, the mole rat of Eurasian grasslands, the pocket gophers in North America. Only Australia lacks a running herbivorous mammal to match the bison of the Northern Hemisphere, the pronghorns of North America, the pampas deer and South American camels, the wild horses of Asia and zebras of Africa, and the antelopes of both Africa and Asia. The Northern Hemisphere has no large flightless grazing bird equivalent to the ostrich, the emu, and the rhea of the Southern.

The Short Grasslands of North America

Between the 100th meridian and the foothills of the Rocky Mountains, the rainfall varies from about 10 inches annually in the north to 25 in the south. About four-fifths of the precipitation comes during the growing season, between April and September, and matches the temperature in each latitude in terms of the moisture stress on plants. The climate is right for short-rooted, low-growing grasses which benefit while there is moisture above the calcified hardpan at the bottom of the B horizon in the soil. Below this level are neither roots nor water.

The native perennial grasses are mostly sod-formers, particularly blue grama (*Bouteloua gracilis*) which grows 6 inches high, and buffalo grass (*Buchloe dactyloides*) which generally is shorter, to 4 inches or less. Mixed among them usually are western wheatgrass (*Agropyron smithii*), 2 to 3 inches tall, and the seedlings of several kinds of woody shrubs.

The shrub seedlings and the wheatgrass can be kept under control only if the animals of the plains and an occasional fire maintain the dominance of the blue grama and the buffalograss. If no grazers keep the

wheatgrass cut, reducing its chance to produce seeds, the wheatgrass forms so firm a sod that it becomes dominant. But if too many grazers keep all of the grasses short, the shrubs grow up and take over. Then a fire is needed to clear the land for grass again.

Until somewhat more than a century ago, this was bison country. Nomadic herds of these big animals, totalling perhaps 50 million head, wandered from mid-Alberta and Saskatchewan to the Texas panhandle, and adjacent Oklahoma and New Mexico. Captain William Clark, while exploring the West in 1804 on the famous Lewis and Clark Expedition, noticed that the bison were never alone:

> I observe near all large gangues of Buffalow wolves and when the Buffalow move these animals follow, and feed on those that are killed by accident or those that are too pore or fat to keep up with the gangue.

Pronghorns, which are not true antelopes but animals that shed an outer horny covering from their permanent branching horns, encountered the same hazards as they roamed the Great Plains in large numbers. Generally they relied upon their impressive speed to escape from wolves. But in a small group with young, surrounded by wolves, maternal solicitude and the instinct of the males to protect the others often let the predators kill them all.

The welfare of bison and pronghorns depended not only on grass and wolves but also on the short-tailed ground squirrels known as prairie dogs. No other denizens on the grasslands so regularly destroyed the seedlings and roots of drought-resistant shrubs, such as sagebrush, that could have replaced the grasses

and starved the grazers. The rodents, in turn, benefitted by frequent passage of the herds. The big animals trampled the vegetation, as well as grazed upon it, and greatly reduced the number of plants in clumps taller than about 6 inches, behind which a coyote, a wolf, a badger, or other predator could approach a townful of prairie dogs. Within about 100 feet of the doorways of their intercommunicating burrows, the prairie dogs kept the vegetation cut short. But they made no attempt to conceal the burrow systems that almost endlessly pocked the plains wherever bison and pronghorns came often. Nor did the prairie dogs evict the small owls and rattlesnakes that came to shelter in their burrows by day.

At night the owls and rattlesnakes emerged to prey on mice. By day, badgers and grizzly bears often tried to dig out the prairie dogs. Often a coyote watched while a badger dug, ready to catch a fleeing rodent before the badger did. Badgers, coyotes, and hawks acting separately caught grasshoppers. They also destroyed many ground-nesting birds, which found little shelter on the plains but were attracted by the abundance of insects and seeds there.

To make the Great Plains safe for domestic cattle and sheep, settlers eliminated the bison, most of the 30,000 Indians who hunted them, the wolves, and the grizzly bears. The swift pronghorns largely ran away, heading into the desert and the foothills. The grass was kept short instead by man's livestock.

The ranchers strove for larger profits by overstocking the range, and the domestic animals prevented normal reproduction of the palatable grasses. This favored rough plants, such as Russian thistle and tumbleweed, or unpalatable ones, such as goldenrod and asters, or poisonous kinds, such as locoweed and the introduced Klamath weed (*Hypericum*, a St. John's wort). Since coyotes, badgers, and eagles were found devouring the carcass of an occasional calf or lamb, programs to eliminate them seemed necessary. But with few predators left to reduce the numbers of mice, grasshoppers,

and prairie dogs, these tended to reach epidemic status and to compete for food with the livestock. One new program of control after another appeared essential to keep the ecosystem biased in favor of domestic animals and the grass to feed them.

Prairie dogs multiplied prodigiously for a while. Under the impression that a cow would eat whatever a prairie dog might, the ranchers instituted campaigns to destroy all prairie dogs. They spread poisoned grain as bait, which killed also the eagles and other scavengers that ate dead prairie dogs. They fumigated the burrows so successfully that the black-footed ferret, which had become a specialist living almost exclusively on prairie dogs in their burrow systems, diminished in numbers. Conservationists now list the ferret among the rare and endangered species in America. Without prairie dogs to clip off the woody plants, the range became vulnerable to invasion by harsh shrubs, particularly sagebrush (*Artemisia tridentata*).

No one seemed aware that this new history in the New World was repeating old history in the Old. On the great grasslands of Eurasia over a far longer period, less vigorous efforts led to a progressive reduction in the proportion of forage grasses and an increase in the prevalence of a different *Artemisia* (*A. absinthium*, known as wormwood) and other harsh native shrubs that livestock ignore as much as possible. Known as "brush invasion," it is a sign of excessive exploitation.

Constant maintenance has become part of the price of so largely replacing the native plants and animals of the Great Plains with introduced kinds. The artificial web of life has almost none of the natural feedback that could provide stability. Undisturbed land is so scarce that the prairie chickens—the last gallinaceous birds anywhere of the genus *Typanuchus* (the "drum necks")—are on the list of endangered species.

Only in the northern part of the great grasslands—chiefly in Canada—is a token of the former wildlife close to its original abundance. There, in the so-called pothole country, low places left by the receding glaciers 10,000 years ago remain as small lakes, marshes, and wetlands in which waterfowl congregate to feed and reproduce all summer. Agriculturalists who might like to fill in these lowlands and transform them into productive fields for grain and forage are generally dissuaded from doing so by private corporations such as Ducks Unlimited, and by government agencies that recognize the recreational values in an abundance of waterfowl. Instead of serving as an outdoor factory for cereals and beef, these lands are a "duck factory." Taxes on firearms and ammunition go a long way toward preserving the habitat for the waterfowl and whatever other wildlife can benefit from the sanctuary.

Sagebrush Versus Grasses

From the intermontane valleys of southern British Columbia southward through Idaho and western Wyoming, to a long triangle along the boundary between Nevada and California, the climate is slightly more arid than on the Great Plains. The carrying capacity of the land is correspondingly less, and overstocking with domestic animals easier. Selective action of the grazers, eating the grasses and letting the harsh shrubs grow, and measures that have been successful in controlling wildfires, have both favored the spread of sagebrush. Today it appears to be an endless stand. although

When the matted roots of native, drought-resistant grasses in a semiarid land are destroyed by cultivation and replaced only by the poorly adapted roots of commercial cultigens, nothing remains to prevent the wind from carrying away the soil during a drought. Decades or centuries pass before a man-made desert or "dust-bowl" of this kind can recover.

actually it is interspersed with lupines and some lesser plants. The lupine flowers lend the color to the famous "purple" sage, the spicily aromatic leaves of which are actually a silvery gray and the flowers an inconspicuous yellowish white.

Formerly this was grassland, over which fires swept every few years and kept the sagebrush low. The lupines regenerated quickly, sharing the nitrogenous nutrients from their root nodules with the fire- and drought-resistant grasses. These were principally feathergrasses and needlegrasses (*Aristida*), which grow in compact tufts. Between the tufts the surface of the soil often remained almost bare of vegetation, although underground the grass roots spread out shallowly, ready to capture whatever rain sank in.

Where the sagebrush has now so largely displaced the grasses, the total precipitation amounts to between 5 and 10 inches annually. Less comes in the north and at low elevations (4,000 feet) and more in the south and higher up (to about 8,000 feet). It is still the home of the jack rabbit and the sage grouse, whose young are able when born or hatched to follow their parent and find shelter beside grass tufts or under a sagebrush. They gain concealment when the sagebrush multiplies, but lose out in the variety of inconspicuous plants from which they might get food.

In the most arid parts, the sagebrush country resembles the bunchgrass deserts of New Mexico, Arizona, and adjacent Mexico. Both have prickly pear and cholla cactus (*Opuntia*) and yucca. But the grasses are different and so are almost all of the surviving animals.

Forest Versus Grasslands

Between the 100th and the 90th meridian in North America, the territory on which the Plains Indians hunted bison included an anomalous grassland almost 500 miles wide. It extended from Manitoba through eastern Minnesota, Missouri, Arkansas and to western Alabama. It grew as a tall-grass prairie where a forest would have been had not the Indians repeatedly set fire to the grasses along the forest edge and killed back the trees. The soil below the tall-grass prairie showed no calcification because rainfall and drainage combined to keep the C horizon wet, below a permanent water table. It was dominated by just two common grasses: the native big bluestem (*Andropogon furcatus*) and the introduced little bluestem (or prairie beardgrass, *A. scoparius*), which spread at amazing speed once it reached the Western Hemisphere. By late summer in some places, these grasses rose to a height of 9 to 12 feet, hiding from view not only the introduced cattle but also any man on horseback pushing through it.

Both of these grasses begin growing in April and continue until September. They achieve almost a third of their annual productivity in June, with only slightly lesser amounts in May and July. But their differences in water relations show during a normal year in that little bluestem is absent from the moister soil at the bottom of each gully ("draw") while accounting for about 70 per cent of the various plants on hill tops. Big bluestem may constitute 97 per cent of the flora in the gully, but be less numerous than the members of the daisy family, the legumes, and other nongrass plants (forbs) on the crests. After a series of dry years, little bluestem is found with increasing frequency on the lower slopes, just as after a series of wet years, big bluestem displaces its competitors progressively higher up.

Neither forbs nor fire are tolerated in most parts of North America where formerly the tall-grass prairie grew. The land is now the "corn belt," and produces crops of greater value than other cereal grains. Nor would it return to tall-grass prairie if abandoned. The climate is still suitable for a temperate deciduous forest. Without human interference, the forest would reinstate itself through an orderly succession if given a century or two.

The Winter Grasslands

A different schedule of precipitation supports a natural grassland in the rolling hills of eastern Washington and Oregon into Idaho, and the broad San Joaquin and San Fernando Valleys of California. The snows and rains come during the cool to cold winters, when the temperature is too low for plants to grow. Summers are hot and dry. To this climate, and to summer fires, a few of the native perennial bunch grasses, particularly fescues (*Festuca*), bluegrass (*Poa*) and a feathergrass (*Stipa pulchra*) are well adapted. They remain dormant all summer, then respond to the first fall rains by extending a few leaves and some flower buds. Dormant again during the cold spells in winter, they show activity whenever a warming of the weather lets them use the moisture and the light under the thin cover of snow. In spring they shoot up, produce their seeds and store enough food in their roots to last through the long summer.

Most of these lands are now occupied by introduced annual grasses, such as Eurasian wild oats (*Avena fatua*) or thatch grass (*Bromus tectorum*) or winter wheat. Just as in corresponding parts of Canada and the U.S.S.R., farmers plow the land and seed it in the late fall, then watch for the "gentle flush of green" as their one crop of the year sprouts after the rains arrive. Enough snow must fall to keep the wheat from freezing to death. In good years it survives and attains a height of 12 to 15 inches by the time it is ready to harvest in May. Good practice calls for dropping the straw left by the threshing machinery on the field to protect the soil until the time comes to plow it under and reseed.

The changes that man has made in grasslands and savannas have mostly been gambles toward harvesting a single crop while making the area drier and less productive than the climate would suggest. Often the gambles have failed, and the grassland has transformed into actual desert. More will be so transformed unless programs for land use can be adjusted in relation to weather cycles that extend over several decades. In this most recently-occupied of the terrestrial environments, the energy relations and nutrient requirements are now understood well enough to allow intelligent planning for "perpetual" use on a scale exceeding the average generation time of the human species.

17 The living deserts

The arid lands of the temperate and tropic zones are called deserts because people desert them, not because they lack native plants and animals. People cannot follow a civilized way of life there without bringing in water from beyond the desert boundaries or raising water from the depths of the subsoil. But a fascinating array of organisms show adaptive features that have evolved over the millennia, allowing them to cope with the almost chronic shortage of moisture. Most of the animals among them are small or nocturnal, and easy to overlook. Few of the plants attain conspicuous dimensions, and these few may give the landscape a monotonous appearance.

None of the world's deserts is centered near the Equator. Instead, they are situated for the most part in the warmest part of the temperate zones. There, on both sides of the tropical zone, the great convection currents in the atmosphere produce belts of high pressure. In each belt, air that has been cleared of moisture by cold at high altitudes continually descends, growing warm with increased pressure as it reaches the earth's surface, and correspondingly ready to absorb water. The dry air could pick up and carry off perhaps 80 inches of precipitation annually if it were available. Where the actual precipitation is 20 inches or less on the average, a desert is inevitable.

According to statistics compiled by the United Nations, about a seventh of the world's land area is outright desert. An equal amount is listed as semiarid, although its bunchgrasses are so unproductive that they can support no domestic livestock. They yield no crops except in the unpredictable wet years that come a few times each century.

The largest desert, with 3.5 million square miles, is the Sahara. Stretching across the

In Roman times, productive grain fields spread across Libya near Tripoli, where now the camels plod through drifting dunes. Attempts in 1951–1952 to anchor the shifting sand by planting grasses failed, but since nomads and their livestock were not fenced out of the test area, little of scientific importance can be learned from the pattern of surviving grasses.

Until 1926, when these twisted remains of a plank road, laid in 1915, were replaced by pavement, travel between Yuma, Arizona, and Holtville, California, was an ordeal for the intrepid. Shifting sands of the Algonones dunes were totally unsuitable for the narrow wheels and hard tires of the first automobiles. They cut to pieces the very first road —a covering of brush. Now Holtville is bypassed and motorists cross the sands on Interstate Highway 8—a distance of nearly 20 miles—in less than 15 minutes.

whole 3,200-mile width of North Africa, it is separated only by the Red Sea from another 1.0 million square miles of Arabian desert in the adjacent peninsula. About a tenth of the Sahara and a third of the Arabian are sandy, rising into a succession of shifting dunes.

The biggest dunes, to more than 700 feet high, are in the smallest of Asia's deserts—the 150,000 square miles of Iranian desert. Larger deserts are the 750,000 square miles in Turkestan, to the north and east of the Caspian Sea; 230,000 square miles in western India (the Great Indian Desert or Thar) and 200,000 square miles of the Takla Makan in western China, which grade into the semiarid Gobi of southern Mongolia.

In Australia, the 1.3-million square miles of desert occupy almost half of the continent, extending nearly to the coast in the northwest and south. Yet they are not quite so arid as the Namib Desert of South West Africa, or the Kalahari of Botswana, in southern Africa, each about 200,000 square miles in extent. South America has a counterpart of the Namib in the Atacama Desert along the Pacific coast from northern Chile into Peru. Although the smallest (140,000 square miles) desert, it is the driest, with less than half an inch of rainfall annually. A Patagonian desert, occupying 260,000 square miles in Argentina lies east of the Andes in the rain shadow cast by prevailing winds.

We are more familiar with the subdivisions of the 500,000-square-mile desert in the American Southwest. Part is in the rain shadow of the Sierra Nevada and Cascade Range, extending across the Great Basin toward the Rocky Mountains in southern Nevada and western Utah. Southward, it merges with the Mojave Desert in southeastern California, and with the Arizona-Sonoran Desert which extends two blunt fingers into Mexico—into Baja California to the west of the Gulf of California and into Sonora to the east. It continues across southern Arizona into southwestern New Mexico, adjoining there the northward extensions of the Chihuahuan Desert, which reach also into southwestern Texas from the main area to the east of Mexico's Sierra Madre Occidental. The plant life in these North American deserts resemble to a surprising extent that of the Patagonian Desert.

In all of these deserts and semideserts, the plants have far more light available than they can use in photosynthesis. Water is the prime limiting factor, and its delivery follows no schedule. A single storm may bring the complete supply for 2 to 5 years, generally at a rate that vastly exceeds the absorptive capacity of the ground. Draining over the surface, the water causes erosion before sinking in or evaporating. Water that does enter the soil tends to rise again by capillarity and be lost too, leaving the soil spaces clogged by salt crystals. Rivers that flow into the desert from adjacent mountains tend to vanish too, some into underground channels.

With so little water vapor in the air above the desert, the solar radiation arrives almost unopposed, baking the soil surface to a firm, water-repellent crust. But the same clear air facilitates radiational cooling at night. Desert temperatures often fall low enough for heavy dew to form on shiny surfaces. The dew evaporates again immediately after dawn. Yet, for a few hours, the relative humidity close to the ground approaches saturation. It is then that many of the desert animals emerge from underground hideaways, and forage in the open for any edible plant materials, for prey, and for carrion.

Lizards, snakes, and carnivorous invertebrates tend to have two periods of activity. They emerge at sundown when the soil surface grows cool and continue active until the night's chill and darkness interfere with their efficiency. Just before dawn they may reappear, exposing themselves to the first rays of the sun until warm enough to move quickly. As the sun gets higher and hotter, they retire underground for the day.

Rodents and other animals that burrow 10 inches or more below the soil surface reach a stratum of virtually constant temperature due to the thermal inertia of the overlying earth. In their burrows at a comfortable 85°F, the air retains most of the moisture from exhaled breath. Challenged by neither heat nor cold, the burrower enjoys the equivalent of air-conditioning. Metabolic needs can be minimal for most of each 24 hours, reducing the need for both food and water. Kangaroo rats in the American Southwest, like the gerbils of North African and Asiatic deserts, actually regain some of the water they use to humidify air entering their lungs. They gather into fur-lined dry cheek pouches the desiccated seeds and fruits they find at night on the desert surface, and store these trophies underground for a week or two. By then the moisture content of the food has increased from perhaps 5 to 40 or 50 per cent, as actual water that the rodent can salvage during digestion.

These same rodents have spectacularly efficient kidneys, which excrete a urine that is almost saturated through reabsorption of water from the distal tubules. At the same time, they show in their behavior an ability to choose foods with high carbohydrate content and low concentrations of fats and proteins. This substantially reduces the amount of water needed for digestion and excretion, generally to the point where the animal has no need during its entire life to seek drinking water. It gets along on "metabolic water" instead.

Ephemeral Life

Fully 40 per cent of the flora may consist of annuals whose seeds can remain dormant for decades, yet sprout if thoroughly wetted for a few days in succession. After a good rain, these ephemeral plants take root, grow quickly but extend only a small rosette of simple leaves before producing flowers. Many of them will set seed without being pollinated, but make use of pollinators if these arrive in time. Within a few weeks the parent plant withers while its fruits ripen. A month after the rain came, the seeds may have been dispersed and the desert floor appear empty of vegetation where a carpet of flower spread previously. Generally the dead plants dry out too quickly for decomposition to return nutrients to the soil. The wind blows away the crisp fragments, crumbling them to dust. So little humus accumulates that the soil rarely shows any A horizon.

In the American Southwest and in some Australian deserts, pollinators do make their appearance soon after a local storm has dropped its moisture. Bees, wasps, and many birds seem to detect the dark cloud in the distance and head for it. They gather around puddles as soon as the erosive flow of water has ceased, and remain until after the ephemeral plants have bloomed.

The shallow warm puddles and ponds, although so temporary, soon teem with algae and with alga-eating crustaceans such as the shield-shrimp (*Triops*) whose

drought-resistant eggs have remained dormant for months or even years. The birds eat some of these animals, but the others grow rapidly, mature, mate, and provide another generation of eggs to carry on their kind. As the water evaporates, the active crustaceans die. The algae leave armored, desiccation-proof, fertilized egg cells. Dusty drought returns, with virtually no evidence that it has ever been relieved.

Life on the Installment Plan

After a rain moistens the soil an inch or two below the surface, many of the woody perennial plants in the desert put out new root hairs and capture as much as possible of the water. Although the stem of each plant may be many feet from its nearest

(Overleaf) Many desert plants put out fresh leaves within a week after a rain but the leaves shrivel as soon as drought returns. In some seasons the stems of the ocotillo (Fouquiera splendens) bear clusters of orange flowers (left). The only other genus in this highly specialized family (Fouquieraceae) has a single species: Idria columnaris, known as the boojum tree (right). Its thick stems bear many short side branches and serve for water storage.

neighbor, its root system spreads out horizontally in concealment and competes with adjacent root systems for space in which to probe for moisture. By comparison, the top of the plant is small—reducing the surface through which evaporative loss is inevitable. This reduction is furthered as soon as the water underground can no longer supply the needs of the newly-active plant. It lets its root hairs wither, die, and decompose. It drops its leaves, but not before new buds have formed—ready to repeat this sequence after the next rain whether it is next month or next year.

Snails in the desert and amphibians such as spadefoot toads live on the same installment plan. They react to moisture by becoming active, feeding while there is food available. Around the deeper pools, the male toads croak excitedly summoning mates who have eggs ready to lay in bead-like strings. In a few days, tadpoles hatch and begin feeding on the algae. At about the same time, young snails appear from smaller eggs that the adult snails have deposited in the water. Eating and growing at top speed, these aquatic animals must reach a critical size if they are to survive in dormancy, with reserves of food and moisture adequate to last until the next rain lets them progress another step toward maturity. Before the pond quite dries out, the tadpoles and adult toads prepare for drought by burrowing downward through the muddy bottom. They seal themselves in mudlined cells a foot or so below the desert surface. The snails may remain in view, each cemented firmly to the woody stem of some shrub an inch or two above the ground.

Unexpected proof of the tolerable duration of dormancy was obtained early in World War II, when bomb damage to the roof and glass top of a display case at the British Museum (Natural History) let rain reach some desert snails on exhibit. When the curators arrived to repair the damage, they found their snails creeping about, each one still with a label affixed to its shell, recording that it had been collected—apparently dead and too desiccated to need preserving—more than 20 years earlier!

Storing the Water

In many deserts, the most conspicuous plants are those that hoard into thick stems or leaves the water they absorb after a rain. They are drought-resisters, rather than drought-avoiders. In the Arizona-Sonoran desert, the giants are saguaro cacti as much as 50 feet tall, sparsely branching and with a shallow root system extending for a radius almost equal to the height. One such cactus may contain more than 10 tons of water, expanding as it absorbs and contracting as the water is used in photosynthesis or lost to the dry air. The lengthwise ribbing of these cacti provides accordion-like pleating for this change in volume, as well as projecting surfaces armed with sharp spines that ward off thirsty animals. Barrel cacti have a similar construction, in a shorter plant. Both have glutinous materials that further protect the water store from animals that could enter between the spines.

Desert environments of southern Africa, although too remote from those of the New World to have native plants in common, have nevertheless selectively allowed the survival of vegetation with matching adaptations in vegetation of unrelated families. This convergent development is particularly conspicuous in members of the genus *Euphorbia* of the spurge family (Euphorbiaceae). Several species closely resemble

Although the knoblike fruits on a prickly pear (top) are often brightly colored and sweet, the surface may be armed with fine spines, showing their cactus origin. A barrel cactus (center) may hold enough water to last it a year or two. The pleating allows for expansion and contraction. Below, an elf owl peers from the doorway that a Gila woodpecker cut while excavating a nest in the stem of a big saguaro cactus.

the treelike cacti, such as the branching organ-pipe cactus of the Sonoran desert, but gain their protection from animals through a toxic milky juice instead of spines. Some of these euphorbias are so well protected in this way that indigenous people have learned to make arrow-poison from the milky sap. Poisonous resinoids of other kinds confer an adaptive value on the milk of cactus-like members of the milkweed family (Asclepiadaceae), especially the strange carrion-flowers (*Stapelia*).

The century plants of America, of which the genus *Agave* in the amaryllis family includes fully 300 species, hoard starches as well as water in their long, thick, wax-covered, and sharp-pointed leaves. These organs arise in a basal whorl close to the ground, in a pattern of growth that has its counterpart in the nearly 200 different kinds of African aloes, in genus *Aloe* of the lily family (Liliaceae). "Hen-and-chickens" represent a much smaller version of storage leaves in a rosette, the leaves being broader, blunter, and tightly grouped—an analogous product of evolution in the stonecrop family (Crassulaceae). This type of plant was named for those native to semiarid Mediterranean coasts, which extend new plants ("offsets") at the ends of short runners from under and around a mature one. The same habit of growth and succulence has allowed other members of the same family to colonize arid deserts. One subfamily includes low-growing species of *Crassula* in deserts of southern Africa, while another subfamily is represented by *Dudleya* species from Arizona to Baja California.

Most succulents are killed if chilled far below the freezing point. Ice needles form inside their cells and puncture their membranes. Seemingly for this reason the cacti and other water-hoarding plants are

mostly missing from deserts that get cold in winter. Outlying populations of cacti, such as prickly pear (*Opuntia*) in southern Canada, survive where the annual precipitation is greater and includes at least a deep blanket of snow over the plants during the frigid months.

The water-hoarding plants may be common in hot deserts and conspicuous in the way that scattered trees are on a grassy savanna. But their presence becomes almost insignificant if they are compared with hard-leaved woody shrubs in terms of the total extent of their canopy, or the total basal area where the plants emerge from the soil, or even the proportion of the desert floor in which their hidden roots are spread.

Harsh Shrubs in a Dry Land

Judged by criteria of root area, basal area, and proportion of the total canopy, as was first suggested by the American ecologist R. F. Daubenmire, the true dominants in most deserts are woody shrubs. Although their hard stems conceal little water-storage tissue above ground, they are drought-resisters. Generally their leaves are small, leathery, able to curl and protect the stomata in the under surface or to drop off, leaving the branch bare when evaporative stress becomes excessive. The stiff petiole may remain, discouraging many browsing animals from eating the twig. Sharp spines of many kinds serve similarly. Often the twigs and foliage contain bad-tasting or poisonous substances that repel herbivores.

In regions where cattle or native grazers and browsers keep the grasses and forbs short and scanty, unable to carry wild fire under the harsh shrubs, these woody plants form a dense stand to which the names chaparral and thornscrub are applied. Often the soil surface below and between the shrubs is bare, although the hidden roots form a dense network and are able at all times to absorb cutoff water.

Sagebrush in North America and wormwood in Eurasia, both members of the genus *Artemisia* in the daisy family (Compositae), are the dominant shrubs in cool northern deserts. Growing two to three feet tall, they seem to form a canopy of silvery gray-green leaves on stiff branching stems. Viewed at close range, other plants with the same growth form are found

mixed among the artemisias. Most of them are members of the goosefoot family (Chenopodiaceae). Those best known in the desert areas of North America are greasewood (*Sarcobatus vermiculatus*), winter fat (*Eurotia lanata*), spiny hopsage (*Grayia spinosa*), and saltbush (*Atriplex*—particularly *A. canescens*, which is called shadscale).

The other three North American deserts more often are dominated by a different composite, called bur sage (*Franseria dumosa*), and by creosote bush (*Larrea tridentata*), which is an excellent binder of loose sand. The Mohave desert has sagebrush too, but its most outstanding plants are the Joshua trees (*Yucca arborescens*), which are tall, branching, grotesque members of the lily family (Liliaceae) found in no other part of the world.

The Arizona-Sonoran desert is equally distinguished by its saguaro cacti (*Cereus giganteus*) and organ-pipe cacti (*C. marginatus*). On higher slopes it supports ocotillo (or cat's-claw, or coachwhip, *Fouquiera splendens*), whose long, slender, spiny, woody stems stand leafless most of the year. They produce new leaves and terminal clusters of flowers after each rain. Low places, such as the "washes" formed by erosion by storm water, commonly become almost solid stands of native smoke trees (*Cotinus americanus*) and introduced tamarisk (*Tamarix*). The Arizona-Sonoran desert has the greatest development of ephemeral annuals.

The Chihuahuan desert is the home of barrel cacti (*Ferocactus*) and the strange boojum tree (*Idria columnaris*) which is related closely to ocotillo. Boojum trees

resemble inverted carrots, to 30 feet tall, topped after a rain by a little cluster of leaves and white flowers which soon complete their activities and drop off. The composite family is represented by guayule (*Parthenium argentatum*), which resembles a dandelion but has a milky juice from which rubber can be made. Yuccas are common, as is a shrub of the spurge family known as candelilla (*Euphorbia antisyphilitica*). Dunes are often topped by a squaw-bush (*Rhus trilobatus*), while valley floors are tufted by a foxtail grass (*Setaria*).

In these deserts and in others all over the world, shrubs and low trees of the pea family (Leguminosae) are particularly numerous. They grow best on alluvial fans, where rivers from nearby mountains flow into the desert and disappear, or where storm water arrives at intervals only to vanish quickly by sinking in and evaporating. Arid America has its mesquite (*Prosopis*) and palo verde (*Cercidium* and *Parkinsonia*), whose roots may extend downward as much as 100 feet and reach moisture in deep drainage channels even during prolonged droughts. The margins of the Sahara and eastward to India form the range of camelthorns (*Alhagi*). Licorice (*Glycyrrhiza*) and acacias (*Acacia*) grow in both the New World and the Old. North Africa has contributed a herbaceous relative with remarkable roots, which is now widely planted in semiarid lands to yield a hay crop known as alfalfa or lucerne (*Medicago sativa*); it remains green and growing long after native plants with shorter roots have died or become dormant. Except for these few deep-rooted types, which are known as phreatophytes ("well plants"), the desert ecosystem is extremely shallow and its ecological niches correspondingly limited.

Animals in the Food Web on the Desert

The abundant small seeds of the ephemeral annual plants and the larger parts of the perennial kinds of vegetation provide a reserve of food for desert animals that are able to find and use the often-poisonous plant materials. Adaptive radiation among the insects has produced many specialists and these, although small in average size, become the food base for carnivores in considerable variety. Scorpions and hunting spiders and centipedes hunt out the insects. So do some of the lizards and many of the desert birds. Other reptiles, such as the chuckwalla and the desert tortoise of the American Southwest, chew on the plants, and eat even the spiniest of the cacti to get sustenance, nourishment, and water.

Often the interactions among the living things bring gains as well as losses. The activities of woodpeckers and other insectivorous birds may significantly reduce the rate at which insects attack plants. Yet woodpeckers commonly open the bark of desert vegetation and let valuable moisture escape. The Gila woodpecker of Arizona makes a practice of hollowing out a nest cavity 8 to 10 inches high in the trunk of some big saguaro cactus. Working quickly, the bird removes the soggy inner tissue faster than the plant can exude a protective coat of sticky sap—material that would almost certainly immobilize and drown an insect that bored into the succulent stem. The cavity produced by the bird becomes lined by an extremely hard layer of callus tissue, which blocks loss of moisture and entrance of disease organisms. Even after a saguaro has died, fallen, and decomposed, the callus tissue remains on the ground for years. Prospectors who found these curious objects could not imagine their origin, and named them "desert shoes."

The nest hole of a woodpecker offers shelter to other birds when the woodpecker abandons it. One of the most picturesque is the sparrow-sized elf owl, smallest of the world's owls. This little bird often can be seen standing in the doorway of the cavity at sundown, waiting for the light to fade before flying out to feed. The elf owl is a specialist in finding and disarming scorpions on the desert floor. It seizes its victim by the tail, snips off the venomous stinger, and swallows the rest of the scorpion in one piece.

Desert mice eat insects as well as fruits and seeds. In turn, many of them fall prey to desert kit foxes, ringtailed cats, bobcats, coyotes, and the various pit-vipers that hunt in complete darkness. One of the rattlesnakes, called the sidewinder (*Crotalus cerastus*), seems specially adapted to life in sandy deserts. It travels sidewise, throwing its body into helical coils that progress across the most yielding of surfaces, even up the sloping sides of sand dunes.

By transporting fruits and seeds to storage centers, animals of many kinds inadvertently disseminate live reproductive parts of the desert vegetation. The pack rat of the American Southwest carefully drags whole segments of cacti (such as "jumping chollas"—*Opuntia* species) and piles them into mounds as much as 15 feet across, as refuges that no large predator can penetrate. Often one or more of the cactus segments develops adventitious roots after a heavy rain, and vegetatively initiates the growth of a new plant.

Pollination requires the services of insects, birds, or bats during the day or two when a desert plant opens its flowers. A failure of pollinators to arrive on time was suggested a few years ago as the probable reason for the slow decrease noted in the number of saguaro cacti in the Saguaro National Monument near Tucson, Arizona. Certainly new young plants were not rising to take the places of old ones that died and fell. The failure in reproduction has become serious not only in the Monument area, which was established to perpetuate a particularly fine grove of these unique plants, but also in many other parts of their range in the Arizona-Sonoran desert. Tests quickly confirmed that each saguaro is self-sterile, and requires cross-pollination to set seed. Honeybees, which were introduced into the area about a century ago, can accomplish cross-pollination if they are numerous enough. But careful observation showed that the natural pollinators are nectar-feeding bats (*Leptonycteris nivalis*) and day-flying western white-winged doves (*Zenaida asiatica*), which visit the saguaros in considerable numbers during the flowering season. The bats visited also some of the century plants (*Agave*) that ordinarily are regarded as being pollinated principally by hummingbirds.

When pollination was shown not to be a limiting factor in the reproduction of saguaros, other stages in the life cycle of the plant were investigated. The cause proved to be destruction of saguaro seeds by desert mice and seedlings by these same and other herbivores in the desert. All through the 20th century, the populations of these first-order consumers has been unusually high because the number of wide-ranging predators—especially coyotes—has been kept down by men trying to raise domestic livestock close to desert country.

Providing protection for saguaros alone is fairly easy. Rodent-proof enclosures of wire mesh can be fastened to the ground as sanctuaries for individual saguaro seeds and seedlings. The task of maintaining these enclosures for years can be reduced by starting the plants in a nursery and transplanting them when they reach 5 to 8 years of age and a height of a few inches. Thereafter their spines and sticky sap enable them to resist attack. Unfortunately, neither program for the benefit of saguaros does anything to improve the balance between growth and destruction for other desert plants, which suffer similarly from the abnormally large number of small herbivores.

Without human interference, the population of predators on a desert is always small, in keeping with the low productivity per square mile enforced by the meager supply of water. The kit foxes and ringtailed cats (cacomistles) of the American Southwest must range widely to find both food and mates, and be extremely sensitive to scents and sounds in the desert night. Ecologists see an additional benefit in the oversized ears of these predators and of prey such as jack rabbits. Not only do these organs follow Allen's rule and provide extra surface from which heat is radiated, but they also gather in sounds from greater distance in a biome where distance is particularly plentiful.

Ranchers in semiarid areas of the American Southwest have developed a rule to follow in relation to carrying capacity. The land is cattle country if it will support one cow or steer to the square mile. Areas of lower carrying capacity are relegated to sheep. Since neither sheep nor beef animals of breeds with high economic importance have appreciable ability to defend themselves or their young from predators, the success of men using the desert fringes as cow country or sheep country depends upon elimination of all predators that might interfere.

Along the edges of deserts on other continents, the customs of local people differ more than the ecological situation. To the south of the Sahara, the herdsmen vigorously defend their domestic animals from actual attack by lions and other predators but rarely venture in pursuit of large carnivores except when proving their manhood. To become marriageable a Masai boy must "blood his spear" by killing a lion. Not until disease control among the Masai allowed the number of maturing Masai boys to rise with no matching increase in the number of lions did this practice threaten the lion population. To the north of the Sahara and in Arabia, predators and prey alike are regarded as targets to be killed by whatever means available. But there, too, plant material that the domestic animals do not eat becomes fuel—and no part of the vegetation remains to reproduce itself and match the actual climate with a grassland or a thorn forest. Even the shrubby trees from which frankincense and myrrh can be obtained are burned by the nomads to boil their beverages or warm their tents.

Water in the Deserts

In many places in deserts north of the Equator in Africa, water flows slowly along in porous rocks, following a gradient from the mountains where precipitation comes regularly to springs along the edge of the Mediterranean Sea or in its bottom. Where the topography brings this water within 60 feet or less of the land surface, it is often possible to dig wells by hand and dip up water in quantities adequate to let date palms grow, generally with small plantations of millet and other crop plants in the shade. Formerly it was possible to estimate the human population of an oasis by counting the date palms, for the many products from each tree were needed in primitive living at the rate of one tree per person.

Often the help of a camel let the oasis dweller raise water more efficiently, one skin bagful of liquid after another as the camel pulling the rope to the bag was led back and forth on a ramp close to the well. But when pumps driven by internal combustion engines were introduced and operated for more hours each day, the rate of removal often exceeded the rate of replenishment and the operation reached a state of diminishing returns.

Drilling to great depths and pumping up water that lies far underground has been tried more frequently in arid parts of the American Southwest and Australia, where people are able to afford the equipment. But these underground reserves are not being recharged at a significant rate, and the water level in them drops progressively until it becomes uneconomical to reach. Most of the water reached these reserves at the end of Pleistocene times and is no longer a part of the dynamic water cycle; it is correctly described as "fossil water."

Irrigation schemes, requiring expensive engineering feats to bring water to the deserts and supply the moisture needed for crops of food and fibers, soon cause a change in the desert soil that requires for correction still more water. The liquid diverted from a river contains modest concentrations of dissolved salts, but this concentration rises on the irrigated fields as the water evaporates or is absorbed by the plants. The field tends to become a salt desert, upon which only a few useless, highly adapted plants will grow until some climatic change brings regular rains to leach out the soluble salts. Improved techniques can postpone the day when the soil becomes too salty for crop plants. But the cost of using them consistently gives a reason for considering an alternative use of desert land. It may be cheaper over the long term to cease trying to bring ever more water to the deserts and to cease shipping produce to the distant cities. A preferable program may be to move people out of all cities where the rainfall and temperature are suitable for agriculture, to reclaim the city land for agriculture, and to ship the produce in the opposite direction—to new cities in the desert.

18 Life at its bleakest

Most people think of the polar regions in terms of snow and ice. They hear of the great glacier atop Greenland and the glaciers of Antarctica as being a mile thick, and suppose that this accumulation represents a few years, rather than millennia. It is easy to progress from an inch of snow in winter at latitude 35°, perhaps in North Carolina, a foot of snow at latitude 40°, such as in Pennsylvania, and 10 feet of snow in Montreal and adjacent parts of Quebec just beyond the 45° circle, and then expect the worst in the Arctic Zone. Actually, the precipitation in polar lands at sea level is meager, and almost all in the form of winter snow. Greater accumulations are typical of higher altitudes, rather than high latitudes.

The polar ice is real and understandable. For half of each year, night is longer than day. During the dark hours, the polar lands radiate to space far more energy than they receive through all these months. At winter solstice, the sun fails to rise at all, just as at summer solstice it circles the sky for at least 24 consecutive hours. These phenomena stem from the geometry of the earth as it orbits the sun, steadily rotating on an axis inclined $22\frac{1}{2}°$ to the plane of that orbit. Even in summer, the rays of the sun are so oblique that they are first filtered through an unusual thickness of the atmosphere and then spread over a large area in polar lands, which minimizes the heating effect.

Winds in the polar regions cause so much mixing of the air that the temperature there never falls as low as in some protected parts of the taiga. But usually it is a moist wind in summer and a dry wind in winter, often whipping snow, ice, or sand at high velocity as abrasive tools. No dormant plant can stand before it. Few other living things expose themselves to the icy gales.

Musk oxen huddle, arctic hares and arctic foxes and polar bears and snowy owls all crouch as low as possible in the Arctic. Emperor penguins in the Antarctic stay close together and turn their backs on the wind.

Musk oxen and arctic hares dig through the snow to reach edible vegetation. They uncover the runways of lemmings that find both food and security in the shallow blanket of snow. When lemmings emerge into the open and hares move freely, they often become prey for owls and foxes. If lemmings are few, the owls migrate southward. The polar bears, like the penguins, get their food from the sea at all seasons.

By remaining low, or being pruned back by the winds in winter, the plants keep their growth centers under the snow blanket protected from the coldest and driest weather. Snow and frost may return, however, all through the summer growing season. For this the plants must be tolerant. Whenever they have liquid water and light, they carry on photosynthesis. This limits their growth to two or three months each year. Fortunately, they can make good use of the long summer days, for photosynthesis is actually more efficient when the light is not too bright.

Recently, the reason was discovered for the paradoxical fact that growth of plants actually improves to the north of the Arctic Circle, although the sun is lower at noon and its heating effect is less. However, nights are shorter too, and the soil does not chill much by radiational cooling so long as the sun is above the horizon. Where the low plants are, the earth stays warmer than a few feet above the ground. This absence of a temperature inversion next to the soil, and the significance of the difference in microclimate there, remained unknown for many

Lichens, mosses (especially
Sphagnum*), sedges, and grasses are
the low-growing vegetation of the
tundra that extends across delta land
from the mouth of the Yukon River
to Kuskokwim Bay, Alaska. They
emphasize the bleakness of the biome
and give no hint of the attractions
offered to migratory waterfowl that
arrive early in the brief summer to
feed and raise their young on the
sudden growth of aquatic plants.*

years because weather records were kept only in relation to instruments at the height of a man's chest. For vegetation that is ankle-high, conditions important to human activities are of little consequence.

In summer, the treeless character of much of this land gives it the appearance of a wet meadow. The greatest expanse of it is in the U.S.S.R., where it is called the tundra. The term has been adopted worldwide, for it applies equally in northern Scandinavia, Iceland, around the southern coasts of Greenland, in northern Canada, and Alaska. In the U.S.S.R. it dips well south of the Arctic Circle, and in northern Quebec and

Labrador to 50° north latitude—almost halfway to the Equator. It continues down the American mountains to Tierra del Fuego, as though to link the north-polar community of life with that around the fringe of Antarctica.

The southern boundary of the Arctic tundra on low land is the edge of the northern coniferous forest (the taiga). Many attempts have been made to learn the basis for the rather abrupt change at timber line The length of the growing season seems especially important, as A. D. Hopkins sought to make quantitative in his bio-climatic law:

Horizontal arrows show the length of the actual growing season, measured in consecutive frost-free days, at four weather stations at different altitudes in Colorado. To at least 10,000 feet elevation, they fulfil predictions based upon Hopkins' bioclimatic law, as projected from the mean frost-free periods in Denver (dotted line), Boulder (white line), and

Longmont (solid line). The highest station (D), with a growing season of more than 50 days rather than just the predicted 27, is exceptional because it is on a tundra ridge immediately east of a higher range, where it warms up rapidly in the summer sun and does not collect cold air at night from any higher areas.

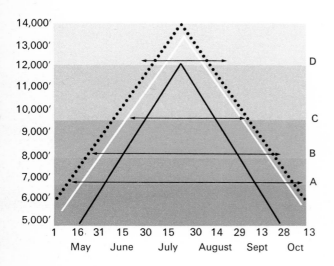

Other conditions being equal, the variation in time of occurrence of a given periodic event in life activity . . . is at the general average rate of 4 days to each degree of latitude, 5 degrees of longitude, and 400 feet of altitude, later northward, eastward and upward in spring and early summer, and the reverse in late summer and autumn.

Hopkins' "periodic events" include the opening of flowers, nesting of migrant birds, and similar activities to which an average date can be assigned. When predictions from this relationship are compared with the actual growing season, as measured in frost-free days, the exceptions stand out. (See diagram above.) They point to "other conditions" that are *not* equal, but that have real significance to living things.

Polar tundras differ considerably from alpine ones. Those in the Arctic overlie a deep soil that is permanently frozen, as a "permafrost" left unchanged since the Ice Age. Rivers cannot easily cut erosion channels into this frozen ground. Meltwater cannot sink downward through it. Consequently, over the large flatlands in which polar lakes, ponds, and bogs abound, the habitat is suitable for water plants, aquatic insects, and waterfowl. Only in a few places, such as along the west coast of Greenland, are the winds dry instead of saturated with water vapor. There the tundra gives way to salt flats, with salt-tolerant plants (halophytes) as a true desert in the Arctic.

By contrast, the alpine tundra has no permafrost. Drainage usually is much better. Even in the Tropical Zone, however, every night on the high mountain slopes is winter, and only sunny days are summer. No mountain in the temperate or tropical zones has a 24-hour day in midsummer, although regularly at every season the sun rises earliest and sets latest on the peaks.

Precipitation, as well as the number of hours of direct sunlight and the temperature, affects the growth of plants both up-slope and toward the poles. Slope is least important in the polar lands, for the summer sun goes around the sky, illuminating the plants for much of the time at an intensity greater than the compensation point of photosynthesis. Under these conditions they grow well. But, on mountains in the temperate zones and between the Tropics, variations in precipitation correspond to compass directions as well as altitude. Tree line is higher on the windward, rainy side than on the leeward, drier side. It occurs around 13,000 feet on slopes under the Tropic of Cancer and the Tropic of Capricorn, but somewhat lower at the drier Equator.

In seeking to explain the actual location of the boundary between evergreen coniferous forest and tundra, whether alpine or polar, the German ecologist W. Köppen tested many possible correlations between altitude and climatic features. His con-

The wingless black orthopteran insect, *Grylloblatta campodeiformis,* has been found only under stones at the edge of the snow line, where it tolerates being frozen for many months each year but dies if exposed to the heat of a human hand. Apparently it has been able to survive in this habitat for hundreds of millions of years, for entomologists have concluded that it is one of the world's most primitive insects.

clusion, that timberline corresponds to whatever altitude follows the isotherm of 50°F (10°C) for the warmest month of the growing season, seems to fit more of the recorded data than any other rule.

The Greening of the Tundra

The Arctic Ocean was almost certainly open water until shortly before the Ice Age. Temperate forests extended to within 7 degrees of the North Pole. As the Pleistocene climate chilled, natural selection favored those living things that were pre-adapted to tolerate the cold. No one knows where this tundra biota originated, for most of the species in it are circumpolar. Many are found also southward along mountain chains as far as alpine Mexico. They must have formed a shifting band between the advancing ice and the forests, colonizing southward as the glaciers spread.

The evidence shows that the survivors colonized northward again as the ice retreated. On isolated peaks, such as the Alps and those in New England, they followed the warming weather up the slope. Now many on the summits or high meadows are absent from lower elevations but virtually identical with species found much farther north. Thus the Presidential Range in New Hampshire and Mount Katahdin in Maine, although only about a mile high, have Labrador tea (*Ledum groenlandicum*), Lapland rosebay (*Rhododendron lapponicum*), and various kinds of sedges and reeds native to low elevations in Labrador and around Hudson Bay. Some of the butterflies show the same distribution.

Tundra, whether in the North or at high altitude, remains a frontier close to the uninhabitable. Nowhere else on earth do living things face such spectacularly changeable weather. A single hour may bring them full sun, pelting hail, sudden snow, violent wind, and then utter calm with full sun again. Flowers are a gamble, and delicate ones impractical. Pollen can be wetted without warning. Chill is likely to prevent insect pollinators from arriving when needed. Cross-pollination cannot be relied upon for bees are few, flies and butterflies erratic in their visiting. Many tundra plants reproduce asexually by bulblets, branching rhizomes, and runners just above the surface of the ground. Some pollinate themselves if no insect arrives. Others have embryos that develop into seeds without pollination at all. In the face of such severe physical factors in the environment, a fixed pattern of inheritance that has been successful seems more important than the variation and adaptability that might be gained by cross breeding. The fixed pattern shows considerable plasticity in the height the plants attain where they chance to find shelter.

Despite their similarities, the plants of the polar tundras do differ from those of high altitudes. They flower early in the season, at or soon after summer solstice, as "long-day" plants. Those from mountains' tundras generally bloom later, as "short-day" plants more like their relatives at low elevations in the temperate zones and between the Tropics. These differences, and the low habit of growth, are retained when the plants are translocated to lower latitudes and lower altitudes, which shows them to be inherited features. The tundra plants retain their tolerance for intense sun, intermittent drought, and winter cold, but not for shade. They may be enjoyed as "rock garden plants," so long as these environmental requirements are met in their new location.

On the polar tundras, the dwarf birch (*Betula nana*) grows prostrate, rising only to a height of perhaps four inches, as the commonest representative of a genus that consists of trees (except in alpine situations, where other species grow). Similar dimensions are characteristic of the northern green alder (*Alnus crispa*), which is peculiar in its genus because it opens its reproductive catkins and its leaves simultaneously; elsewhere alders reproduce late in the season or prior to the appearance of leaves. The nitrogen-fixing bacteria in nodules on alder roots add a scarce nutrient to the shallow soil, stimulating the growth of other plants nearby. The evergreen ones are mostly members of the heath family (Ericaceae), such as cranberries, blueberries, bilberries, and bell-heather. Cottongrass (*Eriophorum*), which grows commonly around pools, is a member of the sedge family (Cyperaceae). It may be rooted in soil, or in a floating mat of peat moss spreading over a pool or an especially wet area of tundra.

Animals of the Polar Tundra

The arctic tundra was once the home of vast herds of reindeer in Eurasia and of caribou (both *Rangifer*) in North America. Those in the Old World were exterminated, except for domesticated animals herded by Lapplanders. The caribou in the New World continued until about a century ago to follow irregular migratory routes between the tundra in summer and the nearest taiga in winter. The evergreen coniferous trees gave them the shelter they needed from the winds, and hindered them little in pawing through the snow to reach reindeer moss and other slow-growing plants. (Caribou is an Indian word signifying scraping or pawing.)

Any increase in the numbers of caribou soon exceeded the carrying capacity of the lichens in winter. Animals that showed signs of malnutrition were eliminated by the attendant packs of wolves, restoring the population to ecological balance. The wolves took the diseased and the crippled animals too, staying close, watching, ready for an opportunity, all year. But healthy, well-fed caribou had little to fear from these predators. The bot flies that followed the herds in summer were far more of a nuisance. So were the black flies and mosquitoes that appeared in thirsty hordes soon after the beginning of the growing season.

The caribou herds shrank when Eskimos and other hunters began to use firearms in place of more primitive weapons. The animals became still fewer when manmade fires destroyed much of the lichen range. This change deprived wolves of their principal prey, and their numbers decreased. Now they must compete with arctic foxes, weasels, and snowy owls for arctic hares, lemmings, and other fresh meat. On this basis about 100 square miles of tundra are needed to support each family of wolves—the parents and about four young.

For all of these vertebrate animals, winter meals are the limiting feature. Populations reach their lowest ebb in spring. Then, almost suddenly, the days lengthen. Open water appears. Plants grow. Migrant birds of many kinds arrive. There is more available to eat than animals to eat it.

The sparrows and longspurs that nest on the tundra, and the waterfowl that breed in enormous numbers around the lakes, ponds, and pools, show a minimum of wariness, even when a predator is close by. Each carnivore spends most of its summer standing or lying around, waiting for its last meal to be digested before getting another. There is no need to go hunting, for potential victims are on all sides. Yet being well nourished does not lead to rapid reproduction of the predators. Litters tend to be small, which reduces the amount of predation. The genetic heritage that regulates the potential for population growth seems well adjusted to the physical environment that lets so few individuals survive the winter.

The only abundant nonmigratory animals are the insects, particularly the blackflies and mosquitoes. Breeding in the pools, small streams, and even in wet peat moss, they emerge in unbelievable numbers, and the females seek eagerly for blood. All summer long the Eskimos continue wearing their fur hoods to protect themselves against these insects—not for warmth, for on the long days the temperature often reaches the 60's Fahrenheit.

Poles Apart

No such pests inhabit the antarctic region, although it bears a symmetrical relationship to the arctic in relation to sun and seasons. Whatever growing season the antarctic has alternates with that in the North. Yet except for the most migratory of birds—the arctic

terns and some tube-nosed birds (shearwaters and albatrosses) of the Southern Hemisphere—no animals regularly benefit from the long days near one pole and then those near the other.

The polar climates actually have little in common, because virtually all of the area south of the Antarctic Circle is high land surrounded completely by open seas, whereas most of the area north of the Arctic Circle is ocean nearly encompassed by broad continents. Essentially all of the Arctic Ocean lies north of 80° latitude, yet its ice cover breaks up each summer because of heat received from the Gulf Stream. The antarctic continent receives no similar supplement to the energy supplied annually by direct sun, and remains far colder.

Romantic tales of the "Roaring Forties" emphasize the frequent gales from the west that circle the planet between 40° south latitude and the coast of Antarctica. Only Tasmania, part of New Zealand, and South America south of Valdivia (Chile) and Bahia Blanca (Argentina) extend into the path of these winds. Except for a few small islands, the rest is open water to the Antarctic Circle. On Antarctica itself, part of the Palmer Peninsula and of the shore of Wilkes Land project into the South Temperate Zone. Technically, all of this windswept water is temperate too. But even in southern summer, it has little heat to share with the antarctic shores.

Something of the climatic differences can be seen in the fact that the southernmost trees in the world grow on South America at about 54 degrees south latitude, whereas in Scandinavia and northwestern North America, forests extend north of the Arctic Circle. Only three species of vascular plants maintain themselves on Antarctica. Two are grasses (*Deschampsia*), and the

other an herb (*Colobanthus*) of the pink family (Caryophyllaceae). By contrast, the arctic tundras have a flora with more than 1,000 species of flowering plants, many of the 800 in North America being represented in Eurasia as well. The algae, lichens, and mosses of Antarctica are more varied, but are of kinds found in the temperate highlands of the Southern Hemisphere as well.

The resident fauna of the southern continent is correspondingly limited. The conspicuous ones are all visitors that get their nourishment directly or indirectly from offshore waters. Penguins nest near the coasts. Skuas prey on penguin eggs and chicks, and nest close by. Seals haul out to bear their young and to breed. About 25 different kinds of parasitic insects sometimes get left ashore, but cannot be considered antarctic animals: biting lice and parasitic mites on the birds, and sucking lice on the seals.

Of true terrestrial animals, Antarctica has about 24 species of springtail insects and biting mites, plus a single kind of wingless midge (*Belgica antarctica*) whose larval stages are spent eating algae in brackish coastal ponds. The largest aquatic animals on the continent are similarly in the size range between 4 and 5 millimeters—less than a quarter of an inch—when adult. They are fairy shrimps (*Branchinecta granulosa*), which swim inverted in meltwater pools, feeding on algae. The other freshwater animals are smaller: unicellular protozoans, flatworms, roundworms, water bears (tardigrades), and rotifers. At least one kind of rotifer inhabits the brackish ponds. Most of these small animals are of wide distribution, being disseminated in a desiccated state by wind. They tolerate being repeatedly wetted and dried, frozen and thawed, and for food find enough algae with similar tolerances.

The cold seas offshore are far richer in dissolved nutrients, in diatoms and other unicellular phytoplankton, and in animals large enough to see easily. The essential intermediaries, however, are the red shrimp-like krill crustaceans (*Euphausia superba*) nearly three inches in length. They sweep up plankton and convert the nourishment into food suitable for small fishes, squids, sea birds such as arctic terns and little Adélie penguins, crabeater seals, and whalebone whales.

The variety of life is not great, but the numbers of individuals and general productivity are enormous. This is interpreted to mean that Antarctica has been isolated from the other continents and its offshore waters from more northern oceans to a major extent for perhaps 70 million years. All but about 10 per cent of the fishes under and among the pack ice are of a single endemic order (Nototheniiformes). Its families are astonishingly diverse, for their members have apparently become adapted to occupy a wide range of ecological niches which, in waters elsewhere, would be populated by equally distinctive fishes of many other orders. Among the most peculiar fish in the antarctic area are the so-called ice fishes (Chaenichthyidae), several species of which seem unique among vertebrate animals in lacking red cells in their circulating blood. Apparently their metabolic rate in the cold water is so slow that they suffer no insufficiency of oxygen reaching their tissues.

The large fishes feed on small fishes and squids. So do the commonest seal (the Weddell), the rare Ross seal, and the largest of the penguins (the emperor). These warm-blooded swimmers, as well as the Adélie penguins and crabeater seals, are preyed upon by leopard seals, females

Extremes of adaptive tolerance are
required in animals of the polar
regions. A source of energy and
nutrients must be within reach
during both the long summer days
and the long winter nights. The
polar bears, which weigh 700 to 900
pounds at maturity, manage to find
enough to eat by remaining close to
open water of the Atlantic Ocean and
connecting bays. There, a wealth of
plankton and organic matter is
maintained in all seasons by the Gulf
Stream, supporting a rich bottom
fauna and large numbers of fishes,
squids, and seals. An expert
swimmer, the polar bear can catch
what it needs. Often a female bears
her young in a shelter dug from the
snow on drifting cakes of ice far
from land (top, right). In the
Antarctic, by contrast, the largest
land animals are emperor penguins
(Aptenodytes forsteri) *standing four
feet tall and weighing about 75
pounds (bottom, right). These birds
reverse the usual reproduction
pattern by climbing out onto the ice
shelf in early winter and incubating
their eggs or brooding their young
despite the darkness and gales. One
parent tends the young while the
other goes to the coast to feed on
fishes and squids. Here too the
supply is rich in all seasons due to
upwelling bottom water and the
circumpolar current known as the
West Wind Drift. In the Arctic the
short-eared white hares (far right,
above) dig through the shallow snow
cover or find edible buds on shrubs
exposed in narrow valleys. The seals
(far right, below) remain close to a
hole cut through the ice, which gives
them alternately access to sea foods
and to air; in spring the young are
often born on the ice and nursed
there until they can fend for
themselves.*

of which grow to be 13 feet long. The top predators are the wide-ranging killer whales (Orca), which hunt in packs.

Today, it is difficult to be sure from the remnant populations of whalebone whales how regular in the past was the migratory pattern of their ancestors. A good many—perhaps most—of the survivors travel from antarctic waters toward or across the Equator when day length shortens in the Southern Hemisphere. Apparently the huge animals fast while progressing through unproductive tropical oceans, but find suitable food again when they reach the North Atlantic and North Pacific. Arctic whales, such as the white whale (*Beluga*) and the narwhal (*Monodon*), travel a much shorter distance into the North Atlantic during the polar winter, returning as soon as the breakup of the ice and increase in day length allows greater productivity among the phytoplankton. Similarly, in the North Pacific, the California gray whales migrate south close to American and Asiatic coasts, giving birth and breeding in sheltered lagoons of Baja California and northern Japan before the end of the year.

Exploitation of the animals that make use of the productivity in antarctic waters has not yet been developed on the basis of perpetual yield. Perhaps it is only natural that men who journey into these gale-swept areas to harvest its products should want to take all they can on each long trip without regard to the future of the resource. For a while the rewards were great from slaughtering the southern fur seals (*Arcto-cephalus*) where they haul out on New Zealand coasts and those of many isolated islands. To find them, men explored the southern oceans despite the hazards from frequent violent storms and the difficulties in landing on the rocky shores. Gradually,

however, the market for the pelts diminished through competition from Alaskan fur seal skins, which are of higher quality. Cessation of sealing in the Southern Hemisphere came none too soon, for the seals had almost been exterminated. They are now making a recovery, and limited harvesting under strict control is being tried.

Attempts to manage pelagic whaling in the southern oceans failed from lack of an enforcement agency and a realistic approach toward matching annual quotas to the rates of whale reproduction. By the 1960's, most of the valuable whales seemed on the way to extinction. One nation after another ceased to send whaling vessels to the Southern Hemisphere, for lack of whales to catch. Whether the surviving populations of large species are now too depleted to benefit from this respite remains to be discovered.

The Bleak Peaks

The alpine tundras are almost certainly far older than their polar counterparts. Flowering plants had mountains to colonize soon after they evolved, for the Cenozoic era began with major mountain building over much of the world. By then each continent had its peculiar share of the new plant families. Today these differences are reflected in the distinctive alpine floras of mountains in Latin America, the West Indies, Africa, southern Asia, New Guinea and Australia, New Zealand and the Pacific islands.

Wind endangers life on the slopes far more than in the high latitudes. The greatest wind velocity on record—231 miles per hour—was recorded from the summit of Mount Washington, New Hampshire, just before the gale smashed the gauge and carried it away. Most kinds of alpine plants

The whitebark pine (Pinus albicaulis), *an alpine tree of western American mountains, becomes a dwarfed, wind-pruned shrub at timberline (left). Cotton grass* (Erio-phorum) *is an arctic sedge, seen conspicuously in wet tundra such as lake shores near Mount McKinley in Alaska (right) and in cold bogs farther south. Its tufts of white hairs around its seeds earn it the name of bog cotton.*

gain protection from this violence by taking fullest advantage of the frictional slowing of the air next to the ground. Many are shaped like rounded cushions, with all of their foliage at a single level, and the flowers barely projecting beyond the leaves. Often the foliage is hairy, the transparent hairs interlacing in a way that blocks air movement but admits light and captures radiant heat. A thermometer thrust into one of these cushion plants may register in the 70's or 80's Fahrenheit while the sun warms the surface, although a foot above the ground the air is in the 30's or 40's.

On the high slopes, plants and animals are above the bulk of the earth's atmosphere and exposed to sunlight that is little filtered, containing high and dangerous amounts of ultraviolet. For animals, at least as important a limitation is in the decreased atmospheric pressure, for they need pressure to cause oxygen molecules to enter the blood and tissues. Reduced availability of oxygen (hypoxia) can limit muscular activity. At La Paz, the capital of Bolivia, an ignited match tends to go out for lack of oxygen; the city boasts of being safe at 11,800 feet elevation without a fire department! At 20,000 feet, the partial pressure of oxygen in the atmosphere is only half that at sea level. Atop Mount Everest (29,002 feet) it is less than a third.

In the low latitudes of the temperate zones and between the Tropics, the climate on high mountains shows seasonal variation in rainfall, but little in temperature. Both of these environmental features vary greatly, however, upslope and down, and in relation to prevailing winds. Within a linear distance of a few hundred feet, horizontally or on the incline, changes in the micro-climate cause obvious differences in the kinds of plants and animals present.

At higher latitudes, variations in the length of day and the angle of direct sunlight causes snow fields to extend farther

downward in winter and to retreat upward in summer. Herbaceous plants lie dormant until the snow cover melts, then quickly grow new leaves and continue capturing sunlight as long as the temperature permits. Many live as precariously as the glacier lilies (*Erythronium*) in the western mountains of North America. These small bulbous herbs produce two or three leaves each year and store enough nourishment not only for the next year's growth of leaves but also, during a period of six or seven growing seasons, to raise a slender stalk with a single flower. The plant may die if an animal eats all of its foliage early in the summer. It has too brief a time before winter returns, and such modest reserves that it can barely produce a second set of leaves and carry on enough photosynthesis to survive the winter.

A few of the plants, most notably some sedges, seem able to survive at the upper limit of their ranges so long as they are uncovered by the melting snow every other summer. This intermittent activity is reminiscent in many ways (including the genera of plants represented) with the rapid but brief growth of woodland wild flowers on the soil of a temperate deciduous forest between disappearance of the winter snow and the closing of the leafy canopy of the trees above. Yet the total productivity during the short season may be little more than the amount on high mountain slopes between the Tropics, where the standing crop of vegetation is meager but contributed to all year.

The winter burden presses prostrate the woody plants, such as willows and birches, that grow close to the snow fields. As soon as the cover melts in early summer, they curl upward, putting out foliage and catkins simultaneously. Often the mountain ridges have more woody plants than the sloping valleys because the snow cover is kept thin by the wind and melts sooner, extending the growing season. The valleys not only become choked with snow, but they are kept cooler in summer by the downhill drainage of cold air from the peaks. The bottom of each valley tends to have a boulder-strewn intermittent brook, which is flanked on each side by shrubless tundra known to botanists as a fell-field.

Birds that nest in alpine habitats adjust to the arrival of winter in different ways. The sparrow-sized water pipits of western mountains in North America fly to sea level, and feed along the Pacific coast in company with surf birds that have come this far south from the arctic tundras. The grouselike ptarmigans migrate no more than a few hundred yards downslope to reach woody vegetation whose winter buds are still exposed. These birds change their brown plumage in patches to snow-white, at a pace that matches the arrival of snow areas; in spring they reverse this sequence, maintaining remarkably fine camouflage. Golden eagles, by contrast, must have meat—either fresh or carrion. To get it in winter they travel to lower altitudes and lower latitudes, where currently the land affords far less food for them than it did a few centuries ago.

The larger mammals of high mountains are mostly cud-chewers that can obtain nourishment from exposed vegetation that has dried to hay as well as from fresh plant tissues. Those that climb highest in summer are extraordinarily surefooted. They clamber along precipitous slopes to reach small patches of vegetation where few predators can follow. In winter, they descend as necessary to avoid the deep snow. Their populations have dwindled recently, less

On warm days the Iceland poppy (Papaver nudicaule) raises its delicate flowers on tall supple stalks. It is found above the tundra across Eurasia and northern North America.

because of hunters than starvation during the cold months. When they reach the high slopes and foothills, they find far less food than formerly because domestic sheep have been pastured on these areas all summer and the normal residue of dead tall grasses is gone. This change has affected the American elk (wapiti), which is the counterpart of the Eurasian red deer, the mountain bighorn sheep, and the distinctive mountain goats. The goats are related closely to the European chamois of similar altitudes, and to antelopes, but not to true goats.

In the high Andes and in Tibet, indigenous peoples have developed physiological adaptations that allow them to be surprisingly active despite the relative scarcity of oxygen at altitudes above 10,000 feet. The Andean Amerindians and the Sherpa tribesmen of the Himalayas show an inherited enlargement of the chest, lungs, and heart, as well as proportionately more red blood cells and hemoglobin per cell. These features increase the efficiency of oxygen absorption and transport in the body. In both cultures, local cud-chewing animals have been domesticated: the llama and alpaca from native camels in the Andes, and the yak from wild stock in the Himalayan highlands. The small vicuñas, which are the only other camels at high elevations, have been left free; uncontrolled hunting has sadly depleted their numbers. Reduction in the amount of carrion has led to a similar decrease in the populations of giant condors and smaller scavengers.

Scientists on expeditions to high altitudes are often surprised to find animals living close to and above the snow line. On the slopes of the Himalayas in 1924, Major R. W. G. Hingston with a party of British explorers discovered a number of small, immature jumping spiders at 22,000 feet elevation. This is nearly 2,000 feet higher than the highest known flowering plants in these mountains—a chickweed (*Stellaria decumbens*) forming low cushions in the shelter of the rocks. For many years no one discredited Hingston's idea that the spiders live by cannibalism, eating others of their kind that came upslope on silken parachutes blown from lower elevations. Recently, however, these spiders and others at similar

307

altitudes were found to feed on springtails and flies. The springtails find pollen grains, spores, and other vegetable matter dispersed by wind on the snow and bare gravel slopes, and also fungi growing on the droppings of birds. The fly larvae utilize these same droppings. On the Himalayan high slopes, the birds include migrants (such as bar-headed geese and hoopoes) passing twice each year, and also snow partridges and yellow-billed choughs that scavenge at frequent intervals as though unaffected by the meager oxygen supply.

Some lichens live about 22,000 feet. The snowfields themselves are often colored bright pink by large populations of the green alga *Chlamydomonas nivalis*, which has an accessory red pigment as well as chlorophyll. Apparently these microscopic flagellated cells adjust their position in the snow, down in full sunlight, up at night, and a long way up after new snow falls. They propel themselves in thin films of meltwater whenever the temperature reaches the melting point. A similar habit may be followed by the slender, half-inch snow worms (*Mesenchytraeus solifugus*), which are red with hemoglobin, just as their relatives the earthworms are at low elevations. Snow worms are known to eat algae and pollen on the snow, but most of their living requirements remain unknown.

Vagaries of wind, precipitation, and tem-perature at high elevations sometimes trap large numbers of migrating animals whose metabolism can be sustained only well above the freezing point. At irregular intervals, clouds of the grasshoppers known as migratory locusts have settled on snow fields or been forced down there by a storm. Chilled into immobility and death, they may be covered by snow and incorporated into the structure of a glacier. "Grasshopper glaciers" are well known in the Rocky Mountains, where thin layers of deep-frozen insects form horizontal dark bands in the melting face of the ice. Thawing to mushy masses, the carcasses are washed into the outflow stream, adding organic matter that has been in dead storage for many centuries.

The bleakest territory on earth, along the margins of glaciers whose moraine has not yet been colonized by obvious plants, is the unique home of some extremely primitive insects: thysanurans (*Machilanus*, "glacier fleas"), collembolans (springtails), and wingless orthopterans (*Grylloblatta*). These living fossils sustain themselves in a precarious sanctuary where the climate is too severe for competitors. Or the windblown debris may be analogous to the food that was available in Silurian or Devonian times, when their ancestors first moved out upon the land beyond the fringe of shore plants onto an otherwise barren, rocky world.

19 The present address of living things

Most species could live in many other places besides their native habitats. Yet, only a few have become virtually worldwide, and most of them were helped, accidentally or on purpose, in being transported from their original homes to distant lands, by the most cosmopolitan species of all—man. The great preponderance of living things get energy and reproduce where they are found because their ancestors evolved nearby or traveled to such places at some time in the past. The ecologist sees the present location of plants and animals as evidence of past dispersal, colonization, and success in overcoming the resistance of new environments. Success led to population increase and speciation.

The first recognition that the distribution of living things showed geographic patterns came in 1858 with the publication of a scientific analysis of birds of the world. The author, P. L. Sclater, then director of the London Zoo, emphasized the following six major realms, each of which comprises the entire range for certain orders, families, genera, or species of birds:

1. Australia and New Guinea (the Australasian realm)—birds-of-paradise, mound-builders, lyrebirds, emus and cassowaries, and cockatoos;
2. Latin America (the Neotropical realm)—toucans, tinamous, screamers, hoatzins, rheas, and macaws;
3. Africa south of the great deserts (The Ethiopian part of the Paleotropical realm)—guineafowl, secretary birds, touracos, and the ostrich;
4. Southern Asia (the Oriental part of the Paleotropical realm)—crested swifts, leafbirds, flowerpeckers, honeycreepers, and most of the pheasants;
5. Northern Asia, Europe, and North Africa (the Palearctic realm)—is less distinctive, with hedgesparrows as the principal family of indigenous birds; many nesting species, particularly of waterfowl, migrate south of the realm for the winter;
6. North America (the Nearctic realm)—nesting species that escape seasonal food shortage by regular migration, mostly to Latin America.

Sclater's conclusions were well known to that great zoological explorer Alfred Russel Wallace, who sought to establish a zoological boundary between tropical Asia and Sclater's Australia–New Guinea area. He placed it between Bali and the eastern, adjacent island of Lombok, because Bali and the islands to the west, including Borneo, Java, and Sumatra, had tigers like those in Malaya and the rest of Asia, whereas Lombok had no tigers but did have cockatoos like those on islands farther east.

In his monumental two-volume work, *The Geographic Distribution of Animals*, Wallace showed that mammals and certain insects followed the distributional pattern that Sclater had recognized among birds. Later biologists accepted the evidence for what they called "Wallace's Line," but grouped the Philippines and Hawaii with Asia. Actually, Wallace's Line is in a zone of transition between realms.

The need for geological perspective in any interpretation of the geographical distribution of life was soon pointed out by the distinguished Hungarian-American scientist and explorer, Angelo Heilprin. His book, *The Geographic and Geological Distribu-*

For thousands of years the white storks (Ciconia ciconia) *have nested in Europe between the Baltic Sea and the Mediterranean, close to the wetlands in which they can find the frogs and insects they feed on. People have built nest platforms for* the storks on rooftops and tall poles, but recently the number of the birds has dwindled due to the reduction in their feeding territories and nest sites, hazards such as antennae, and perhaps agricultural poisons.

tion of Animals (1887), introduced this new concept, and, in addition, grouped together North America with Europe, North Africa, and northern Asia as the Holarctic realm, each of which showed more similarities than differences. This circumpolar area included all of the world's musk oxen, moose, caribou, walrus, wolverines, weasels, and ermines, lemmings, beavers, varying hares and ptarmigans, and most of the tailed amphibians.

It was not for another twenty-five years that a thorough understanding of the geography of and the ecological relationships between plants was realized. Schimper showed in his book *Plant Geography on a Physiological Basis* (1898) that the geographical distribution of plants corresponded to the various realms of animals. Schimper also added two new plant "kingdoms": the antarctic kingdom, including remote islands in southern oceans, New Zealand, the tip of Tierra del Fuego, and the Falkland Islands; and secondly, the Cape region of South Africa.

Age and Area

In trying to account for the geographical distribution of life, J. C. Willis compared plants that seemed native to large areas ("wides") with those found only in restricted regions (endemics). He summarized his discoveries in New Zealand:

Willis suspected that this phenomenon of inverse proportion of endemic and direct proportion of wides species was due to differences in the length of time that species had had in which to disperse themselves: the longer the time, the greater the area occupied. In 1918 he expressed his views in a book, *Age and Area*. To avoid examples that could be dismissed as co-incidences, he limited himself to groups containing at least 10 closely allied species. From the outset he admitted that the pattern of dispersal of species could be "enormously modified by the presence of barriers such as seas, rivers, mountains, changes of climate from one region to the next, or other ecological boundaries, and the like, also by the action of man, and by other causes." To test Willis' generalization, both careful analysis of old data and new research were required. This research led to gains in a scientific understanding of the present geographical distribution of life and also the patterns of dispersal. Unfortunately, one variable, which Willis overlooked or underestimated, disrupts the application of his rule more than all of the factors he listed: individuals and species vary enormously in their ability to reach and colonize new areas. Whereas Willis treated all organisms as essentially similar in this respect.

Range in miles	Number of endemic species	Number of wides
1–160	296	30
161–400	190	38
401–640	184	53
641–880	120	77
881–1080	112	201

Continental Drift

The similarities among living things on the separate continents and islands of the Southern Hemisphere have long challenged scientists. Could chance alone, or long routes of dispersal by way of land bridges (such as between Siberia and Alaska and between Mexico and Colombia) account for the presence of porcupines in South America, North America, and Africa but not Eurasia, for camels in the Andes and across the Northern fringes of the Sahara from Morocco to the Arabian desert and the Caucasus? Or was there a direct land route between continents in the Southern Hemisphere—a lost Atlantis or alleged Gondwanaland?

The German meteorologist Alfred Wegener became so fascinated by these and similar questions that he compiled an impressive list of similarities between the plants and animals of eastern South America and western Africa, and in 1924, pointed out how perfectly the coasts of these continents could be fitted together. He suggested that they had once been joined and had separated as the South Atlantic Ocean formed. Until the water barrier became too great to cross, the plants and animals of the two southern continents could have easily dispersed. Additional evidence from paleontology has supported an extension of this theory. The seed fern *Glossopteris* of Permian times and several associated plants are known from fossils found in South America, Africa, India, Australia, and Antarctica, all of which can be fitted together to form a single supercontinent that might well be called Pangaea ("all land").

Late in the 1950's, oceanographers became intensely interested in a mid-Atlantic ridge. Extending from near Iceland almost to Antarctica, it curved in the same pattern shown by the continental coastlines. On each side of the ridge, rock samples showed matching features and gradually increasing age, as though rocks were being formed at the ridge and spreading in all directions from it as new sea floor. Geologists began to visualize vast convection currents in the molten matter of the earth's interior, making the Atlantic Ocean progressively wider, separating the Old World from the New. Since geologists have generally accepted the concept of sea-floor spreading, they have observed also the progressive disappearance of ocean bottom down the major deep trenches in the Indo-Pacific, like a pattern of vast conveyor belts going into the earth's interior. Computers have been programmed to find the most perfect fit between continents bounded by their continental shelves.

Close as the new computations come to the hypothesis that Wegener advanced, they offer less than had been hoped to account for the present distribution of life. All measurements lead to identification of the Mesozoic era as the principal time of continental movements. By then, land plants and animals were well established, but the groups that are dominant today had not yet appeared.

The terrestrial vegetation that could have spread among the continental masses of Pangaea before the continental drift would be psilopsids, horsetails, clubmosses, ferns, and gymnosperms as the only seed plants. Probably lichens, mosses, liverworts, and hornworts were there too, although no fossils have yet been found. Both invertebrates and freshwater fishes show patterns of dispersal prior to continental drift. However, the amphibians and reptiles from the Triassic are not those that survived to the present. Instead, new ones evolved after the continents separated. In the Jurassic, none of the present orders of birds or mammals had yet evolved, none of the insects with a pupal stage, nor the flowering plants.

Just prior to the Mesozoic, perhaps 250 million years ago, during Permian times, the great mass of Pangaea seems to have been separated into a northern super-continent, "Laurasia," and a southern one, "Gondwanaland," by the formation of an equatorial sea called the Tethys, opening to the east but ending where the North American component of Laurasia had just begun to draw away from northwestern Africa, creating the first identifiable part of the Atlantic Ocean. The expansion of sea floor between South America and Africa is dated in mid-Jurassic about 150 million years ago. The Indian Ocean probably had its origin about 110 million years ago in early Cretaceous times, as Antarctica, with Australia and New Zealand attached, swung away from Africa in one direction, India and Ceylon in another, leaving Africa itself with only Madagascar and Arabia. Later in the same period, about 80 million years ago, New Zealand separated from Antarctica. The separation of Europe from North America by the expansion of the North Atlantic Ocean is perhaps 60 million years old, and provided a broad shallow seaway between Norway and Greenland. Apparently, it became a land bridge for a while in later periods, and restricted the flow of Atlantic water into the Arctic Ocean.

Until the Eocene, about 40 million years ago, the Tethys seems to have extended eastward over Asia Minor, connecting north of India with waters along the coast of Southeast Asia. The Iberian peninsula then swung south, opening up the Bay of Biscay on one side and narrowing the Strait of Gibraltar on the other. Australia could have broken away from Antarctica at this time. Apparently, it rotated almost a quarter of a turn as it drifted eastward. Eurasia may have been rotating clockwise while India, with Ceylon trailing behind, moved toward the Asiatic mainland. When this movement ended in the Oligocene period, the eastern end of the Tethys Sea was closed off, thus producing the Gulf of Aden, and raising marine sediments to spectacular heights in Miocene times with the formation of the Himalayas and the plateau country of Tibet.

Reconstructions of the probable positions of the major continental masses have recently been made from evidence of many kinds, in terms of the boundaries shown by the continental shelves rather than where sea level stands at present.

Indications of longitude and latitude are in relation to the present. The dotted portion of the boundary for northern India and southeastern Asia reflects uncertainty because too little geological study has yet been given these regions.

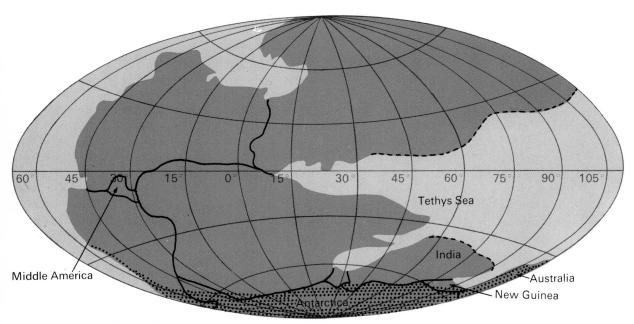

Tethys Sea

India

Australia

New Guinea

Middle America

Antarctica

PANGAEA IN PERMIAN TIMES

250 million years ago

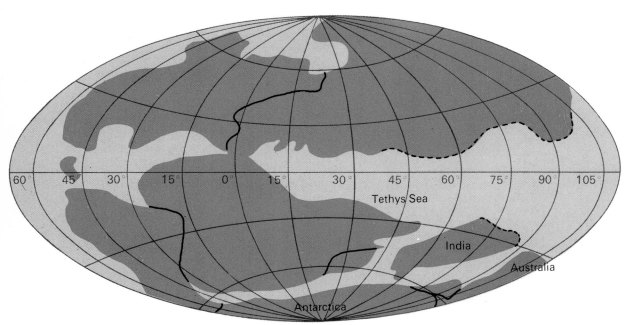

Tethys Sea

India

Australia

Antarctica

LAURASIA AND GONDWANA IN LATE TRIASSIC

185 million years ago

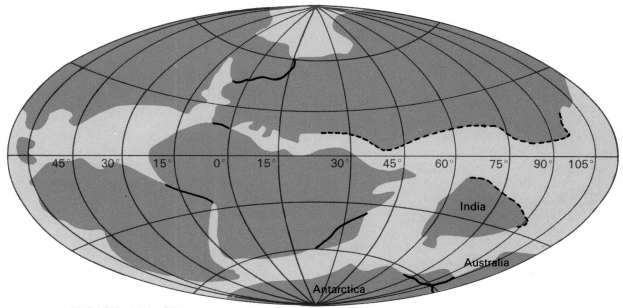

LAURASIA AND GONDWANA
IN LATE JURASSIC

140 million years ago

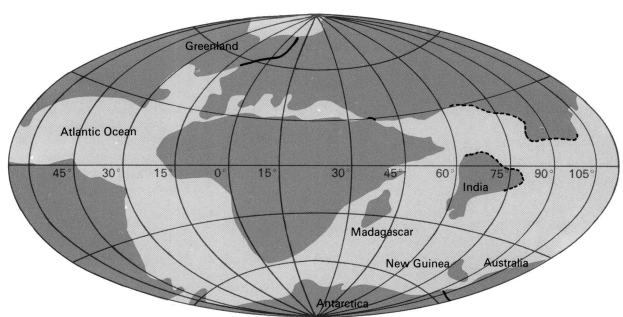

THE CONTINENTAL PATTERN
IN LATE CRETACEOUS

65 million years ago

The Geographic Range of "Living Fossils"

Biologists rely increasingly upon the fossil record, despite its admitted incompleteness, to show which kinds of life are relatively young and hence have not had time to disperse far, and which are relics ("living fossils") from the past, when their ancestors had a broader distribution. The relics are least likely to be discovered in the Holarctic realm, since only small areas of it remained unglaciated during the Ice Age. Essentially all of the biota now found in this realm have reached their present location within the last 15,000 years.

South of the Tropic of Cancer, a biologist has a better chance of discovering these living fossils. Even though their range now may be only a fraction of the spread they showed prior to the Ice Age, it can be interpreted confidently as being derived by progressive constriction. Sometimes the survivors show such poor tolerance that it is hard to comprehend how the organisms persist at all. The velvet worms of phylum Onychophora, for example, accept habitats only where the air is between 90 and 95 per cent relative humidity, and where no contamination ever reaches them. Originally these animals were all placed in the genus *Peripatus*. Now two families of them are recognized. One (the Peripatidae) has a scattered distribution from western Mexico and some islands of the West Indies to southern Brazil, on the equator in Gabon (the sole outpost known in Africa), and in Asia from Burma to Sumatra. The second family (Peripatopsidae) has one group of genera in South Africa and New Guinea, and a different group in South Africa, New Guinea, various parts of Australia (including Tasmania), New Zealand, and southern Chile. All of these can be regarded as descendants of marine ancestors such as the one (*Aysheaia pedunculata*) known from the middle Cambrian of British Columbia. They had 400 million years or more to spread around on land, and may have disappeared from the North Temperate Zone only within the last two million.

The destruction of fossil-bearing rock by the glaciers of the Ice Age affected the more superficial strata representing Cenozoic and Mesozoic times, rather than those from earlier periods. It probably ended forever a paleontologist's chance to learn when the most primitive of vascular plants disappeared from the flora north of the Tropic of Cancer. The earliest psilopsids are fossilized in Devonian strata near Edinburgh, Scotland. Each of these plants is a mere branching stem with distinctive vascular tissue but no roots, leaves, cones or flowers. Supposedly it was once anchored in mud but extended into air. The plants that most resemble it today are two species of *Psilotum* that can be regarded as pantropical. One is present in southernmost Florida; both reach the south island of New Zealand, in the South Temperate Zone. The single species of *Tmesipteris* grows as an epiphyte in wet forests of the same south island, to 45° south latitude, as well as in similar habitats on Tasmania, mainland Australia, and New Guinea, and in the Philippines. Possibly it would grow in the temperate rain forest of Olympic National Park on the Pacific coast of Washington state. That it does not seems due more to its inability to reach this area, where members of the genus may well once have lived.

At the beginning of the Mesozoic era and before continental drift began, the most advanced of vascular plants were the gymnosperms. Today less than 1,000 species of them remain. Of these the members of the cycad order seem least changed since the days of the dinosaurs. *Cycas*, which includes the sago "palm" (*C. circinalis*) of Indonesia and the Philippines, includes 15 species ranging from eastern Africa and Madagascar across tropical Asia and Australia into Polynesia. Two other genera are limited to Africa, two to Australia, and the remainder to tropical America. An exception is the Seminole breadplant (or coontie, *Zamia floridana*), which grows north of the Tropic of Cancer in southern Florida. Fossils show that this distribution is but a remnant from former times, when trees of cycad species comprised major forests in both the northern and southern temperate zones as well as in the tropics.

The ancient order of gymnosperm trees that is now represented only by the maidenhair (*Ginkgo biloba*) apparently evolved during the Mesozoic in the Northern Hemisphere, for fossils of this and other genera, beginning in the Triassic strata of the British Isles, Alaska, and Oregon, are well known. The sole survivor was discovered along the southern border of the Palearctic in Asia, where it was cultivated as a curiosity. More recently the ginkgo has shown incredible tolerance for the polluted air and paved ground of cities in northern Europe and the United States.

Today most gymnosperms are conifers, partly because members of the pine family (Pinaceae) and the cypress family (Cupressaceae) burgeoned forth between the Equator and the arctic tundras through adaptive features that helped them colonize the debris left by retreating glaciers of the Ice Age. These two families account for about two-thirds of the existing conifers. They agree in being "needle-leaved" trees and shrubs; most are evergreens. The cypress family includes not only cone-bearing woody plants but genera such as *Juniperus* whose seeds are borne in fleshy structures that can be called "cones" only by stretching the term beyond customary limits.

To a biologist in the Southern Hemisphere, neither needle-leaved nor cone-bearing seems an appropriate description for familiar members of this order. The principal "conifers" south of the Equator are the 80 kinds of podocarps (family Podocarpaceae), whose evergreen leaves are often broad, and the six different leafless phylloclads (family Phyllocladaceae); both produce seeds with a fleshy coating that attracts animals and leads to seed dispersal. Somewhat similar "fruits" surround the single seeds of yews, of which 19 species in 4 genera are shrubs and trees of the Northern Hemisphere and a single species in a fifth genus represents the family (Taxaceae) in New Caledonia—midway between New Guinea and New Zealand. The Southern Hemisphere does have true cone-bearers in the strange araucarians (family Araucariaceae), such as the monkeypuzzle tree (*Araucaria araucana*) of Chile, the decorative Norfolk Island "pine" (*A. excelsa*) from islands north of New Zealand, and the wonderful kauri "pine" (*Agathis australis*) of New Zealand itself. All of these gymnosperms live where, for one reason or another, flowering plants have not yet displaced them. The gymnosperms were there first, and now are hangers-on as "living fossils."

The Changing of the Guard
Slightly more than halfway through the Mesozoic era, which is now known as the time of fastest continental drift, as well as the Age of Reptiles, the Cretaceous period

During the Paleozoic era, the environment supported at least 456 genera of marine lamp shells (brachiopods) compared to the 68 still surviving; six different species of Platystrophia are shown life size, separated from the matrix of Ordovician rock (top, left), while one common Spirifer is seen embedded in Devonian limestone (top, center). Ammonites were important in coastal seas as shell-bearing cephalopod mollusks from late Silurian until the Cretaceous; remains of Aegoderas capricornus are abundant in some Jurassic strata (bottom, left). The Green River shales of Eocene age in Wyoming are famous for the fossil fishes they contain (bottom, center). The fossilized plants can be even more challenging because the seed habit arose among trees with fernlike leaves in forests or swamps of the late Paleozoic, when giant clubmosses and giant horsetails were numerous. Of the plant fossils above, all from Carboniferous strata in Illinois, the middle left is a stem of a giant horsetail, Calamites, and the one to the right is a seed with whorled bracts of another horsetail, Annularia; in the lower right is a seed, Pachytesta, in its husk; the others are leaves of various seed ferns, such as Alethopteris (lower left corner), Pecopteris (upper right corner) and Neuropteris.

319

began about 135 million years ago. Flowering plants were virtually unknown. Only two of the three orders of mammals that had appeared in Jurassic times were still in existence, but their members were few and inconspicuous in comparison with the dinosaurs. Yet by the end of the Cretaceous, some 70 million years later, the guard had changed. Flowering plants had spread out of Greenland and the northern continents into the southern ones in a succession of unprecedented magnitude. Aided by their newly evolved pollinators, they replaced the older styles of vegetation on most of the good land. The dinosaurs dwindled away. So did one of the two orders of mammals (the Allotheria or Multituberculata), which disappeared from the fossil record after the following period (the Paleocene); the egg-laying platypus and the spiny anteaters of Australia and New Zealand may be descendants of these.

The other order, Triconodonta, became widespread during the Mesozoic. By mid-Cretaceous it gave rise to the pouched marsupials in South America and the placental mammals in Africa, Eurasia, and North America. In South America the marsupials diversified into opossums of many sizes, shrewlike caenolestids and big carnivorous borhyaenids, including some that resembled sabre-tooth cats. Somehow a stock of marsupials spread to Australia and New Guinea, and there founded new lines of many types. But whether this emigration was across open water from South America to Antarctica to Tasmania, or via Central America and North America through Asia, has not yet been learned from the fossil record. In Australia the pioneers found a sanctuary and evolved into molelike burrowers, banded anteaters, pouched mice, native "cats," wolflike car-

nivores, bearlike koalas, bandicoots and possums, phalangers and kangaroos.

Of the placental mammals, a dozen out of about 18 orders that evolved on the northern continents still have survivors. Ancestral lines spread widely, died off locally, leaving the present distribution of insectivores, primates, scaly anteaters, lagomorphs (rabbits and hares), rodents, even-toed hoofed mammals (mostly Eurasian in origin) and odd-toed ones (North American), carnivores, fissipedes (seals and their kin), and less successfully, the flying lemurs (Dermoptera) and the aardvark. Bats flew everywhere. The edentates, such as armadillos and sloths, seem to have originated on Central America where it was an isolated island. The hyraxes, elephants, and sirenians (manatees and dugongs) appear to have had an African origin, and whales to have begun along the southern shore of the Mediterranean Sea.

Active Dispersal of More Modern Life

The population ecologist is always aware of the individuals in each species that find no future in parental territory but that have a chance if they can emigrate and find a suitable ecological niche. The barriers to emigration may be climatic, and change over the millennia. They may be straits and oceans that are inhospitable to a terrestrial organism, or continuous even mountainous land that a marine form of life cannot pass. These limitations, too, are temporary when viewed over thousands or millions of years.

At present, Bering Strait between Asia and Alaska is 56 miles wide and from 60 to 90 feet deep. But during the Eocene period, again in late Miocene to early Pliocene, and for a while during the Pleistocene, this was the site of a forest- and grass-clad plain linking the two continents. If all of the present shallows were above sea level, which is probable, this corridor of land was 1,300 miles broad!

The barrier between marine animals in the eastern Pacific and those in the Caribbean is also new. Today the land is a mere 40 miles broad where the Panama Canal was cut through, and 130 miles wide in Mexico at the low Isthmus of Tehuantepec. Where these links now provide a land bridge between North America and South America, seas washed freely until the Pliocene and again from late in that period until the Pleistocene.

Each of these corridors has permitted two-way traffic, but also served as a selective filter. It has held back some potential colonists while letting others pass. Only cold-tolerant plants, birds, and mammals were close to the Bering Bridge between Asia and Alaska; for cold-sensitive species, even the approaches to the bridge remained intolerable. Horses, camels, perhaps porcupines and freshwater characin fishes spread westward from America into the Old World, while elephants, bison, bears, jays, crows, and chickadees crossed in the opposite direction. The elephants diversified in the New World, and some kept walking across other bridges all the way to Argentina. Crows reached Tierra del Fuego. One kind of bear found a home in the Andes, as the only animal of this type in the Southern Hemisphere. Chickadees, with 42 different races in America compared to 300 kinds and subkinds (called tits) still in Eurasia, have not yet progressed beyond Mexico.

Generally the traffic through a temporary corridor is quite unequal in the two directions. A richer community of life on the one side is likely to include more aggressive species, with special ability to disperse into the territory that is newly accessible to them. The first bridging of the straits between the Americas let many kinds of mammals move south, but few came in the opposite direction. At the time, the southern continent was well stocked with marsupial mammals and edentates. Yet they seemed too specialized to compete successfully with the vigorous newcomers or to take advantage of the land bridge to the north. The placental mammals that went south displaced many of the endemic species, until mostly the opossums and some edentates survived. One of each did reach North America: the "Virginia" opossum, which is notably versatile in its food requirements, prolific, and able to find shelter in many situations, and now common as far as South Dakota and southeastern Canada; and the nine-banded armadillo, which reached the southern United States during the present century.

By contrast, North America had fewer birds to share than South America did. Equatorial parts of the neotropical realm have more kinds of birds than any equal area in the world. As the sea barriers narrowed and disappeared, the tropical and mountain avifauna spread northward. More emigrants found suitable habitats by extending their range a short distance than by going far. The degree to which colonization and separation of ecological niches was achieved during the last 13 million years can be seen from information on nesting species in the chart below.

In Ecuador and Colombia, the nesting species of birds total more than a fifth of all known in the world. Of the 86 families represented, 31 are endemic to this area, an endemism that exceeds twice over that to be found elsewhere. Part of the explanation lies in the rugged landscape, which provides isolation in valleys cut off by mountain barriers. Climatic variations in the last 60 million years have alternately constricted and expanded the useful habitat for birds nesting in these regions, applying natural selection and then providing opportunity for new genetic types to find separate ecological niches. Yet in tropical areas, the population of each species ordinarily remains small. Its constant interaction with so many different neighbors seems to foster rapid evolution.

Ecuador and Colombia have more kinds of birds on New Year's Day than any other part of the world, because the resources in the many habitats are shared by migrant birds from North America, which remain until spring weather is about to recur in distant lands. Barn swallows, kingbirds, scarlet tanagers, rosebreasted grosbeaks, redstarts, sora rails, blue-winged teals and other northern birds compete for food and space so vigorously that the native species suspend their reproduction until the overwintering contingent leaves.

Panama offers most to the bird-lister, because so much variety is concentrated in so little space. Generally the average number of square miles per species gives a good indication of the breadth of the breeding habitat. The figure for Greenland, however, does not fit this pattern since

Colonization and separation of ecological niches by birds

Place	Number of species	Area in square miles	Square miles per species
Ecuador	1,500	109,483	137
Colombia	1,395	439,513	314
Panama	1,100	29,209	26
Guatemala	496	42,042	111
New York State	195	46,576	254
Newfoundland	118	156,185	1,320
Greenland	56	840,000	15,000

only the coastal fringe of the country is suitable for inhabitation, and then only during the extremely short summer.

Island Life Means Isolation

For a land animal to cross open water and reach an island is seldom easy but occasionally possible. On a few occasions, scientists have been able to follow the spread of plants and animals across a barrier and observe their interaction as they tested their tolerances in a new land. The first big opportunity of this kind came in 1883, when a violent volcanic eruption destroyed all life (with the possible exception of earthworms and burrowing insects) on the island of Krakatau. This island is about 25 miles from western Java, where the growing season in the tropical lowlands allows plants and animals to disperse their disseminules almost every day in the year. Within three years after the eruption, blue-green algae had spread over much of the volcanic debris on Krakatau. Botanists found 11 species of ferns and 15 of seed plants. By 1887, another species of fern had arrived and 35 more seed plants. In 1906 the flora included 114 species.

Animals came to Krakatau almost as rapidly. In 1889, bugs and beetles, moths and butterflies, flies and some other insects were collected there. Spiders were numerous, and so were lizards of one species. By 1908 the fauna totalled at least 263 species, including 4 of land snails, 240 of arthropods, 2 of reptiles, and 16 of nesting birds. In 1920, pythons were among the island residents, and the fauna had grown to 573 conspicuous species, including 2 of bats and one of rats.

Probably 90 per cent of the colonists to Krakatau arrived by air. Waves cast others ashore, some coming on rafts blown by winds. Both routes become increasingly chancy with distance and, as the paleontologist George Gaylord Simpson has pointed out, there is no reward for individuals that survive only to within a few miles of a landfall. He compares the arrival of a colonist at a habitable place to the finishing of a race by a horse in a contest. None that falters along the way gets counted. This is "sweepstakes dispersal."

Among the islands of the western Pacific, a progressive series can be recognized, showing that more species crossed the open sea for short distances and few for long jumps. Land turtles seem not to have progressed east of New Guinea. Frogs of several genera reached the Solomon Islands, presumably as passive riders on

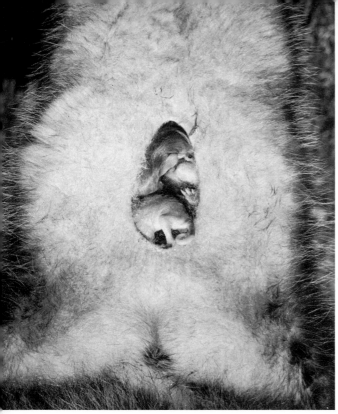

mammals, the eucalyptus trees, and other forms of endemic life. Probably a small contribution came from the south at a time when the Antarctic continent was hospitable at least around its shores. New Zealanders, by contrast, recognize the Antarctic as a significant source of their native living things. Additional colonists arrived (and are still arriving) from Australia to the west, from New Caledonia to the north, and to a much smaller extent from islands and the remote Americas to the east. Within the last few decades, the Australian white heron and a small bird called the white-eye have crossed the Tasman Sea— nearly 1,500 miles of open water—and begun nesting in New Zealand. The buff-backed heron of Africa spread to South America and then to North America at approximately the same time (since 1930) that its range broadened eastward to Australia.

The Hawaiian islands are still more isolated. Prior to the arrival of man, the only native vertebrates were a bat and a number of birds. The bat (*Lasiurus cinereus semotus*) is a subspecies of the widespread American hoary bat, which is a migratory type. The land birds probably represent 14 separate events of colonization, according to the renowned systematist Ernst Mayr of Harvard University. To him the degree of evolutionary change is a reliable measure of the length of time since the colonists arrived. On this basis, the most recent is the black-crowned night heron, which seems indistinguishable on Hawaii from birds of this species in America. Endemic subspecies, which include a stilt, a gallinule, a coot, and a short-eared owl, presumably arrived somewhat earlier, all of these as contributions from the New World. The endemic species—a hawk, a crow, and a

drifting rafts since they are killed by contact with salt water. The same conveyance probably explains the presence of non-swimming snakes, of which several colonized the Solomons and beyond as far as the New Hebrides and Fiji. Some lizards spread to all of these islands and still farther—to Tonga, but not yet to Samoa.

Almost all of the native species on these islands in the western Pacific came originally from the East Indies and tropical Asia. Among the few exceptions with a North American origin is the monarch butterfly (*Danaus plexippus*). In recent years it has spread across the wide Pacific Ocean in the opposite direction. Since it is a strong-flying migrant in North America and often swept for great horizontal distances by warm winds, the possibility of unaided dispersal is open for it. It is similarly possible for individuals in recent times to rest or ride on trans-Pacific ships, and to reach island destinations within decades instead of having to wait millennia for the right combination of weather conditions and time of life history.

Australians, on an island continent, believe that their flora and fauna came chiefly from the north and long ago, giving plenty of time for evolution to produce the outstanding diversity among the marsupial goose—probably all had an American origin but came still earlier.

The endemic genera of land birds on Hawaii include six that belong to families found elsewhere. Dr. Mayr believes that the Hawaiian rail and the Laysan rail (now extinct), both flightless and of unknown origin, came prior to the ancestor of the two genera of honey eaters, which could have come as an immigrant from Australia or New Guinea. Somewhat later, a fly-catcher of an Old World family probably came from Polynesia, and a thrush and a duck from America.

The first land birds to colonize Hawaii were almost certainly ancestral to the present 9 genera and 22 species of honey-creepers in the endemic family Drepaniidae. The pioneer probably was a nectar-feeder of a related family from tropical America. As its descendants multiplied and competed with one another, the adaptations with survival value lay in the direction of altered food requirements. New lines evolved, specializing in insects, fruits, or seeds. This subdivision of a single ecological niche into 22 different ones can now be recognized in different habits, tolerances and structure in the honeycreepers on the major islands in the Hawaiian archipelago.

The volcanic nature of the Hawaiian islands, like that of the Galápagos group, makes difficult any precise estimate of the geological period at which land appeared, providing a place for terrestrial colonists. All of these islands formed in deep parts of the ocean where previously none had appeared. Presumably these events were far more recent than the drifting of the Australian continent into its present position, or the barriers that have kept it isolated from most living things to the north and west.

The diversity of suitable habitats that a new immigrant can use varies tremendously according to the presence or absence of other species. When the ecological niches are essentially filled, the newcomer may be prevented by competition from finding more than a limited scope in the community. When man inadvertently introduced the house mouse in New Zealand, where the only non-marine mammals were two species of bats and a Polynesian rat, the immigrant spread into tussock country, onto the floor of the native bushlands and the new plantations of Monterey pine, into pastures and grain fields. Although the omnivorous mouse overlooked no opportunities to live indoors with man and eat his stored foods, it ceased to be strictly a house mouse. By comparison, when the same species was introduced into the British Isles during the Middle Ages, it encountered 10 different native kinds of mice and voles. Through all the intervening years, the distinctive local rodents have kept the house mouse from spreading into the niches they held before its arrival.

Islands have served as refuges for species that reached them or that became new species under island conditions of isolation. So long as the climate remained tolerable and no new immigrants displaced the island species or destroyed the habitat, they could survive indefinitely with little need to change. The giant land tortoises that long ago reached the Galápagos Islands and also remote volcanic islands in the Indian Ocean maintained themselves over a long time half a world apart but with only specific differences—until sailors began hauling away the reptiles as food, and introduced rats, hogs, and dogs that ate tortoise eggs and young.

Islands become traps for species, which have nowhere to go when environmental conditions change. On the islands of the West Indies alone, more species of mammals and birds have become extinct in the last 500 years than on all of the continents in the past 2,000. Additional species there are now endangered, partly from predators and persecution, but mostly from loss of habitat necessary for their survival.

The Dispersal of Marine Life

Although the oceans are continuous and always have been, the plants and the animals of salt waters meet many barriers to dispersal. The pelagic life in the dark depths has the most uniform environment, and includes many genera found at comparable levels in all oceans. Planktonic organisms show limitations in their tolerances for heat and cold that allow them to disperse widely east and west but not so far north and south. The living things associated with the continental shelves, in the neritic province of the sea, are most influenced by oceanic currents. These carry plankton, larvae, and other disseminules in definite directions according to the pattern of water movements. Winds propel the water. Continents deflect the stream. Where the coastal currents flow toward the Equator from polar regions they maintain an environment suitable for living things that thrive best in cold waters, often far into the Torrid Zone. Conversely, the Gulf Stream and other currents carrying heat polewards enable animals that are best adapted to warm waters to live far from the tropics.

The greatest barrier to marine life in the neritic province is the deep open ocean beyond. This is particularly true of the vast eastern Pacific. Yet the plants and animals near shore between Baja California and the Colombian coast are more like those of the Caribbean Sea than those north or south, because, until recently, there was no land barrier all the way from North America to South America. These understandings of the interplay between ecology and the dynamic geological changes in the continental patterns and intercontinental currents have come only recently. The first comprehensive account, pointing out the distribution of marine animals, was offered in 1935 by Sven Ekman in a book revised and translated from German into English as *Zoogeography of the Sea* (1953). In it, Ekman subdivided the faunas of the neritic province into nine distinctive parts and added two more to recognize the ecological uniqueness of the floating community in the Sargasso Sea east of the Bahama Islands and of the sea floor itself :

1. A Tropical Indo—West Pacific Province, with many reefs and atolls, having sea snakes, blue sea stars, giant *Tridacna* clams, 3 of the 4 kinds of horseshoe crabs, and sea anemones attended by damsel fishes;
2. An American Tropical Coastal Province, with stinging corals, coquina clams (*Donax*), spiny lobsters (*Panulirus*), and the keyhole sand dollar *(Melitta)*;
3. An African Tropical Atlantic Province, with different sand dollars (*Rotula*), but many genera and some species found on both sides of the South Atlantic;
4. A Subtropical Atlantic Province, from Brest in northwestern France along the Atlantic coast to Dakar, plus the saltier Mediterranean Sea and the brackish Black Sea, in all

of which fishermen catch coalfish and whiting (both close kin of cod), plaice (a flatfish), and Norway lobsters (*Nephrops*); the Mediterranean has precious coral and sardines;

5. An Atlantic Boreal Province, along European coasts north of the English Channel and down the North American coast from Labrador to northern Florida, as the territory of eels, Atlantic salmon (*Salar*), the Atlantic lobster (*Homarus*), and a plain sand dollar (*Echinarachnius*);

6. A Pacific Boreal Province from Baja California to Alaska and down the Asian coast to Japan, without eels but with grunions, abalones, sea otters (*Enhydra*), the *Echinarachnius* sand dollar, and coarse kelps;

7. The Arctic Shallows, with capelins, walruses, narwhals, and northern fur seals (*Callorhinus*);

8. The Antarctic Shallows, with other seals (the leopard, crabeater, Ross and Weddell), most penguins, and endemic ice fish;

9. Shorelines with cold antarctic currents—the Humboldt Current to the Galápagos Islands, with albacore and Pacific bonita feeding on the anchovetas, as do Peruvian cormorants and the most northern of penguins; the Benguela Current off South Africa, with different fish and jackass penguins; and waters off southern Australia and New Zealand, with southern fur seals (*Arctocephalus*), fairy penguins and others, and bull kelps;

10. The Sargasso Sea, with two endemic kinds of Sargasso weed that are peculiar in lacking a mode of attachment for their floating fronds, a wonderfully camouflaged angler fish (*Histrio*) and a distinctive pipefish;

11. A seafloor province linking antarctic to arctic waters, cold all the way, with animals that seem unaffected by the great range in pressure, such as most of the world's sea lilies (crinoid echinoderms).

Among these provinces, a few animals of extremely ancient lineage have been able to survive as living fossils. Some, such as the sea lilies, remained so well concealed from human eyes by living on the sea floor that only their fossilized remains in sedimentary rocks had been seen before 1873. Then scientists aboard H.M.S. *Challenger* on a pioneering, 3-year oceanographic cruise, dredged up fresh sea lilies from great depths. The surprises continue. In 1957 from the bottom of the Pacific Ocean off the coast of Mexico, 14,436 feet below sea level, sampling dredges brought the first known survivors (now named *Neopilina galathea*) of a group of mollusks that previously were known only as fossils from the early Paleozoic era. The oldest surviving genera are lamp shells dating from Ordovician times: *Lingula* has many living species in shallow warm waters along Asiatic coasts of the Indian and Pacific Oceans, while *Crania* is found cemented to rocks in the West Indies and some shores of northern Europe. Almost as ancient are the lines of horseshoe crabs, of which four different kinds still represent an ancestry that can be traced to the Silurian period

more than 400 million years ago, and the modern *Latimeria* from moderate depths near Madagascar, where it represents the lobe-fin fishes known otherwise from fossils of Devonian age.

The Indo-Pacific province is by far the largest, extending more than 15,000 miles at the latitude of the Equator from the east coast of Africa to the west coast of South America. Yet over a time measured in millions of years, the sea snakes of the family Hydrophiidae extended their range through most of its warm waters. They are venomous kin of terrestrial cobras, but drift with the plankton, feeding on fish near the sea surface and reproducing ovoviviparously. From a center of evolution in the coral reefs near Malaya, they were carried in tepid waters of the Indian Ocean south along the African coast all the way to Cape Agulhas. This point, rather than the famous Cape of Good Hope, separates them from the cold northbound Benguela Current which flows from antarctic regions along the coast of South West Africa. So far, no temporary cessation of the Benguela or diversion of its flow has allowed the sea snakes to round the corner and enter the South Atlantic. Adaptation to cold water would accomplish the same result, but no one can predict when or if either of these changes will occur.

So far no sea snake is known from Atlantic waters. This is taken as proof that sea snakes colonized the easternmost Pacific only after the land bridge between the Americas developed in the early Pleistocene. The Panama Canal has not breached the barrier for the snakes because they rarely swim against a current and seem intolerant of fresh water. In Panama the man-made facilities include a high-level area (Gatun Lake) filled with fresh

waters from the Chagras River. Water from the lake flows into both oceans through the locks that have been installed to move ships up and down. Without a simultaneous adaptive change toward tolerance for low salinity and toward swimming vigorously through locks and lake, the sea snakes are unlikely to pass through the present inter-ocean waterway. Only the most tolerant of marine life has made this trip; usually attached to a ship or held captive in its ballast tanks during the several days of transit.

Recently, engineers have been studying the feasibility of broadening the present locks to accommodate larger ships and heavier traffic, or of building a new canal entirely at sea level to link the two oceans, and ecologists are trying to predict the consequences. The idea has been set aside because it would make the water in Gatun Lake brackish.

At the Pacific end of any new canal, through Panama or adjacent Colombia, the water is colder, with fewer dissolved nutrients, and nearly 12-foot tides; it has a modest flora and fauna, spiny lobsters

(*Panulirus interruptus*), but no parrot fishes, eels, or sea cows. Along the opposite coast of the isthmus, the lobster (*P. argus*) is of a different kind; there are parrot fishes, eels, sea cows, and a fairly rich fauna and flora in warmer water. The one-foot difference in the level of mean low tide, higher at the Pacific end, would not produce an effective flow from one ocean to the other since the length of the channel will be at least 40 miles. Moreover, so much rain falls on and drains into any low place in this area that any sea-level canal would contain brackish water—never the full salinity to which sea snakes and most other local marine organisms are adjusted. The probability of their dispersal through a new canal seems small, and no ecological catastrophe is anticipated from improvements in the ship channel by methods that are being considered.

The Suez Canal has allowed quite a few kinds of life to spread from the Indian Ocean to the Mediterranean, and some southern animals (such as the green crab, *Carcinus maenas*) of shallow water have used the Cape Cod Canal in Massachusetts to bypass the barrier provided by the cape and colonize the coast as far as Maine. The common periwinkle snail (*Littorina littorea*), which was originally introduced from Europe into Nova Scotia, has found a passage south and has reached New Jersey shores.

Increasingly, the geographic distribution of microbes, plants, and animals is being altered faster than would naturally take place, because of human interference. Within the limits imposed by climate upon terrestrial and freshwater life and by the great currents upon organisms in the seas, the tendency is toward homogenization of the world. The tendency itself is natural and as old as life. Only the rate and the values (for better or worse) that man recognizes in each alteration have changed.

Prior to the advent of *Homo sapiens*, the dispersal of living things was contingent upon fortuitous transport of a viable disseminule, slow climatic changes, or for dramatic geological events such as the eruption of a whole new archipelago of volcanic islands, the immersion of an isthmus, the formation of a land bridge, or the shifting of a continent.

Yet, whether at a slow and natural pace or as rapidly as at present, the sequel to dispersal is competition for ecological niches. Competitive exclusion marks the loser in the contest for local or worldwide extinction, and confers upon the winner the the opportunity to diversify and speciate in the new environment. So long as the changes do not also simplify the ecosystem, life increases the efficiency of its utilization of the resources of its world. Ecologists who have battled to comprehend the complex interactions among living things now realize that the measure of life's complexity is also a measure of life's hold upon the future.

Appendix

Since the middle of the 18th century, when the modern classification of living things began in the binomial nomenclature of Carl von Linné, the number of genera and species known to science has risen steadily. To keep so many categories in a comprehensible array, botanists and zoologists have erected closely matching patterns of categories in ascending sequence above the level of the genus. Genera are grouped into families, those of animals often ending in the letters *idae*, those of plants commonly ending in *aceae*. Families are grouped in orders, such as the Carnivora, the Primates, the Liliales. Orders are grouped in classes, such as the Mammalia and the Angiospermae, often with convenient suborders. Classes are grouped in larger categories, which botanists generally call divisions and zoologists call phyla.

Following the international rules of nomenclature, generic names and those of higher categories are regularly spelled with a capital letter. Generic and specific names are printed in italics or underlined when written. Specialists every year give names to newly discovered subspecies, species, and genera.

With less than 365,000 species of plants recognized by scientific names today, it seems strange that the botanists found need to drastically revise their scheme of classification from class upward so often. The system proposed in 1883 by the German botanist A. W. Eichler showed five divisions, with three (Thallophyta, Bryophyta, and Pteridophyta) as "cryptogams" which lacked seeds, and two (Gymnospermae and Angiospermae) as "phanerogams"—the seed plants. Soon after the beginning of the 20th century, botanists transferred their allegiance to the system of the German botanists H. G. Adolf Engler and Karl A. E. Prantl, with 14 subdivisions:

Schizophyta (blue-green algae and bacteria)
Myxomycetes (slime molds)
Flagellatae (flagellates)
Dinoflagellatae (dinoflagellates)
Heterokontae (yellow-green algae)
Bacillariophyta (diatoms)
Conjugatae (desmids and a few others)
Chlorophyceae (green algae)
Charophyta (stoneworts)
Phaeophyceae (brown algae)
Rhodophyceae (red algae)
Eumycetes (fungi)
Archegoniatae (bryophytes and pteridophytes)
Embryophyta (seed plants)

This, in turn, was supplanted after 1942 by a system proposed by the American botanist Oswald Tippo, with 12 phyla instead of divisions, all but two of them in subkingdom Thallophyta of the plant kingdom, and the final two in subkingdom Embryophyta because of their multicellular sex organs and embryonic development of fertilized egg cells:

Cyanophyta, 1,500 species (blue-green algae)
Euglenophyta, 400 (certain flagellates)
Pyrrophyta, 1,000 (dinoflagellates)
Chrysophyta, 10,000 (diatoms)
Chlorophyta, 6,500 (green algae)
Phaeophyta, 1,500 (brown algae)
Rhodophyta, 3,000 (red algae)
Schizomycophyta, 1,400 (bacteria)
Myxomycophyta, 500 (slime molds)
Eumycophyta, 77,000 (fungi)
Bryophyta, 25,000 (mosses, liverworts, and hornworts)
Tracheophyta, 260,000 (vascular plants)

Since the phylum Tracheophyta includes all of the conspicuous land plants and more than 70 per cent of the species in the plant kingdom, biologists everywhere have needed time to become familiar with its present subdivision into:

Subphylum Psilopsida (including one class, one order, one family, two genera, and three species, two of them whisk-ferns)
Subphylum Lycopsida (with one class, three orders, four genera, and about 742 species of clubmosses and quillworts)
Subphylum Sphenopsida (with one class, one order, one family, one genus, and about 25 species of horsetails)
Subphylum Pteropsida, the "leafy plants"

Pteropsids embrace members of the class Filicineae (the ferns, with about 140 genera and 8,210 species), the class Gymnospermae (the "naked seeded" plants, with about 61 genera and 696 species of cycads, conifers, the ginkgo, and a few others less widely known), and the class Angiospermae (the flowering plants, with seeds enclosed in a fruit). Angiosperms include more than 200,000 species of dicots and at least 50,000 of monocots.

The classification of the animal kingdom reached relative stability prior to 1900, and in 1968 was seen as constituting 24 phyla containing about 1,071,000 named species:

Protozoa	28,350
Mesozoa	50
Porifera (sponges)	4,800
Coelenterata (Cnidaria)	5,300
Ctenophora (comb jellies)	80
Platyhelminthes (flatworms)	12,700
Entoprocta	75
Nemertinea (ribbon worms)	800
Aschelminthes	12,500
Gastrotricha, 170	
Rotatoria (wheel animalcules), 1,500	
Nematoda (roundworms), 10,000	
Nematomorpha, 230	
Kinorhyncha, 100	
Acanthocephala (spiny-headed worms)	500
Priapulida	8
Mollusca (mollusks)	107,250
Sipunculida (peanut worms)	250
Echiurida	150
Annelida (segmented worms)	8,500
Onychophora (velvet worms)	70

Tardigrada (bear animalcules)	350
Pentastomida	65
Arthropoda	838,000
Merostomata, 4	
Arachnida (spiders), 57,000	
Pantopoda (Pycnogonida), 500	
Crustacea, 20,000	
Chilopoda (centipedes), 2,800	
Diplopoda (millipedes), 7,200	
Pauropoda, 380	
Symphyla, 120	
Insecta, 750,000	
Lophophorata	3,750
Hemichordata	80
Echinodermata	6,000
Pogonophora (beard worms)	100
Chaetognatha (arrowworms)	50
Chordata (chordates)	43,000
Tunicata, 1,300	
Cephalochordata, 25	
Vertebrata, 41,700	
Agnatha (lampreys, hagfishes), 50	
Chondrichthyes (sharks and rays), 550	
Osteichthyes (bony fishes), 20,000	
Amphibia (amphibians), 2,500	
Reptilia (reptiles), 6,300	
Aves (birds), 8,600	
Mammalia (mammals), 3,700	

Of these animals, about 847,000 are terrestrial, 167,000 aquatic (mostly marine), and the rest parasitic in animals of the land or waters.

Glossary

Many of the terms used in this book and appearing in magazine articles have special connotations as they apply to the current crisis in our environment. Since even unabridged dictionaries do not include this usage in all instances, we offer short explanations. The great lexicographer Samuel Johnson noted that "definitions are hazardous" and added that "In this work, when it shall be found that much is omitted, let it not be forgotten that much likewise is performed."

Abiogenesis. The production of an organized living entity from nonliving precursors.

Abiotic factor. A nonliving feature of the environment.

Abyssal. Found at a depth of 3,000 feet or more in a marine environment.

Aerobic organism. Any form of life that uses oxygen in respiration.

Aerobiology. The study of the transport of organisms, chiefly as spores, by winds.

Age-specific change. An alteration in function or structure occurring at a definite age in each individual.

Allopatric species. Closely related species whose geographical ranges do not overlap.

Amensalism. An interaction between members of two species in which one inhibits the other from sharing the same resources in the environment.

Ammonification. The formation of ammonia from inorganic substrates by certain bacteria, commonly in soil.

Amplification, biological. An increase in the concentration of a poison as shown at progressively higher trophic levels.

Anadromous. Migrating at spawning time to deposit eggs in waters of a salinity lower than that of the open ocean.

Anaerobic organism. Any form of life that needs no oxygen for respiration.

Antibiotic action. An inhibitory effect on living processes in members of one species due to secretions from another species.

Antibody. A protective substance produced in the blood of an animal in response to the introduction of a foreign protein (an antigen).

Aphotic zone. Those depths in oceans and deep lakes to which no measurable daylight penetrates.

Aquifier. An underground stream carrying water toward the ocean.

Autecology. The study of environmental relationships of organisms belonging to a single species.

Autotrophic organism. Any form of life whose sources of energy and nutrition are independent of any host, prey or organic matter.

Bark wash. Soluble material excreted or deposited on the bark, and transported to the soil by precipitation.

Barophilic organism. Any form of life that tolerates or requires continuously high hydrostatic pressure in its normal aquatic environment.

Basal area. A comparative measure of the space occupied by all members of a plant species from the area of cross section of their stems or trunks above ground.

Bathypelagic organism. Any form of life that swims at great depths in the sea but remote from the bottom.

Behavioral adaptation. A repetitive bodily action by an organism whereby it avoids or minimizes an environmental stress.

Behavioral ecology. The study of the social environmental features of animals as these affect access to resources.

Benthal, benthic. Pertaining to life on or in the floor of an aquatic environment, i.e., associated with the bottom.

Biocidal. Having a lethal effect on a wide spectrum of life, usually the killing of non-target organisms by a pesticide.

Biocoenosis. A European term equivalent to the concept of the ecological community.

Biodegradation. Decomposition of organic compounds through the activity of microbial decomposers, used chiefly regarding synthetic compounds.

Biogenesis. The production of new individual organisms through the reproductive activities of pre-existing, parental organisms.

Biogeochemical cycle. An endless sequential transfer of a chemical element through the bodies of plants and animals, and return to soil, sea or atmosphere.

Biogeography. The study of the geographical distribution of species and of past events that produced these patterns.

Biomass. The quantity of organic substance produced on an area, calculated for a species by multiplying the number of individuals by their average weight.

Biome. A subdivision of terrestrial environments according to the dominant form of plant life in the climax communities permitted by climate.

Biosphere. That portion of the world occupied by living organisms, including all of the hydrosphere as well as parts of both the geosphere and atmosphere.

Biota. The totality of all plants, animals and microbes in an area or a geological period.

Biotic factor. An influence due to an interacting organism, whether of the same or of a different species.

Biotope. A subdivision of a biome, such as the bark of trees in a forest biome, or a sandy (rather than muddy or rocky) area of a beach.

Buffer species. An alternative food resource used by a predator when its preferred prey becomes less available.

Calcification, soil. A natural process common in grasslands where the roots absorb most of the cutoff water, leaving a hardpan of leached lime below them.

Canopy. The horizontal expanse of tree branches bearing foliage.

Capacity, carrying. The finite nutritional resource in an ecological community, ultimately as energy from the producer level flowing to higher trophic levels.

Carcinogen. Any agent capable of inducing abnormal

and uncontrolled growth, producing a cancer in a multicellular animal or plant.

Catadromous. Migrating at spawning time to deposit eggs in waters of full salinity, after a life in waters of low salinity.

Catarrhine monkey. An Old World monkey (family Lasiopygidae) in a subdivision of the primates that includes the anthropoid apes as well.

Chamaephyte. A low shrubby perennial plant with buds between 1 and 10 inches above ground level.

Chlorinated hydrocarbon. A category of synthetic insecticides containing chlorine, hydrogen and carbon, notable for their resistance to biodegradation; chiefly DDT, DDD, Aldrin, Dieldrin, Endrin, Heptachlor, Lindane (BHC), and Toxaphene.

Class. A classification level of organisms encompassing one or more orders.

Climatogram. A graphical representation of the mean values plotted for monthly temperature versus precipitation.

Climax, ecological. The community of life that ends an ecological succession.

Cloud jungle. A special form of tropical rain forest on a mountain slope or low peak, maintained by frequent mists, clouds, and showers.

Cohort. A portion of a species population having similar ages because the members were hatched, born, or germinated in the same year or cluster of years.

Colloid. A state of matter characterized by small particles with distinctive properties resulting from their large surface-to-volume ratio.

Colonization. The establishment of members of a species population beyond former geographical boundaries across a barrier.

Commensalism. The sharing of food resources by members of two unrelated species populations, one of which appears unaffected by the other's gains.

Community, ecological. The interacting individuals of all species populations in an ecosystem.

Community ecology. The study of the interactions of living components of an ecosystem, centered on biotic rather than abiotic factors.

Compensation point. The intensity of light utilizable in photosynthesis that supplies energy at the same rate as the plant is using energy in metabolism.

Competition. Efforts by individuals of the same or different species populations in a community to use the same limited resources.

Cover. A comparative measure of the space occupied by all members of a plant species.

Crop, standing. The biomass that can be supported by the available energy flowing through an ecosystem.

Cryptozoans. The commonly unnoticed animals living in the spaces of the soil.

Cultigen. A domesticated organism from which the human species obtains useful materials such as food or fiber.

Cutoff water. Precipitation that sinks into the ground and progresses through aquifers to the sea.

Decomposer. A member of the trophic level in which energy and nutrients are obtained from organic compounds in remains of dead organisms.

Deep-scattering layer. A stratum in the ocean, sometimes called the "phantom bottom," which reflects sound waves at some depth, usually less than 600 feet.

Dentrification. The release by organisms, usually soil bacteria, of gaseous nitrogen by simplification of nitrogenous compounds.

Deprivation, ecological. A reduction in the diversity of environmental challenges, particularly biotic, due to simplification of the ecological community.

 sensory. A reduction in the diversity of stimuli from the environment to an animal in isolation.

Detritus. Organic matter accumulating on the soil or as a sediment from an aquatic environment.

Diapause. A period of greatly reduced metabolic rate accompanied by dormancy, part way through the growth and maturation process of an insect.

Disseminule. A young organism, newly released by the normal reproductive process, capable of moving or being dispersed away from the parent.

Dominance. The degree of control of a community by members of one or a few species.

 cyclic. A succession in which seral stages lead to the reappearance of a species in a position of temporary dominance time after time.

 species. The ability of members of one species, through adaptive features, to compete successfully for required environmental resources.

Drift, continental. A translocation of major land masses from a single supercontinental area or perhaps two such areas.

 genetic. A change in gene frequencies in a species population that becomes isolated from the rest of the gene pool and that is atypical of the species as a whole.

Drought, physiological. A relative unavailability of water because of rapid drainage or contamination with toxic materials.

Dysphotic zone. A stratum in an aquatic environment where daylight is too dim for photosynthesis but adequate for the orientation of sensitive animals.

Dystrophic lake. A body of fresh water in which dissolved materials inhibit the growth and survival of many kinds of organisms.

Ecosphere. The totality of all ecosystems, hence of all life and the physical environment with which living things interact.

Ecosystem. An ecological community and its physical environment of extent such that the cycling of nutrient materials and energy flow form a closed system.

Ecosystem ecology. The study of energy flow and nutrient recycling in a living community, such as a forest ecosystem or an isolated lake.

334

Ecotone. The boundary between two biomes where normal climatic variations favor first the colonists of one biome and then those of the other.

Edaphology. The study of soil as a habitat for plants, chiefly in terms of the abiotic factors.

Edge effect. The distinctive opportunities afforded along the boundary between two plant communities for animals that can feed in one and take shelter in the other.

Effluent. Wastes that decrease the quality of water, air or soil; usually pollutants discharged from a sewer, chimney, or ditch.

Elfin woodland. A type of tropical rain forest on mountain sides and crests where trees are dwarfed by wind and combinations of temperature and precipitation.

Emigration. The movement of individuals from a species population past former barriers into territories where they may succeed as colonists.

Endemic organism. Any living thing whose known geographic distribution is limited to the area in which its ancestors presumably evolved.

Endophyte. A plant, usually microbial, that lives within another plant as a customary site, whether as a parasite, a commensal, or a mutualistic symbiont.

Environment. The combination of physical and biological features that surround organisms in an ecosystem.

Ephemeral annual. A desert plant that grows quickly from seedling to maturity and then survives only in the seed stage until rain again stimulates germination.

Epiphyte. A plant that grows atop another plant, thereby gaining access to light but without absorbing nourishment from the support.

Equivalent, ecological. An organism whose ecological niche in one ecosystem corresponds closely to that of an unlike organism in another ecosystem.

Ethology. The study of animal behavior as a scientific counterpart to human psychology.

Euphotic zone. Aquatic habitats down to the depth at which too much light energy has been absorbed for photosynthesis to continue.

Euryhaline. Tolerant of a wide range of variation in the salinity of the aquatic environment.

Eurythermal. Tolerant of a wide range of variation in the temperature of the environment.

Eutrophication. The addition of dissolved nutrients to the waters of an oligotrophic lake, rendering it more able to support rapid growth of plants.

Exclusion, competitive. Progressive attainment of dominance by members of one species and simultaneous diminution of a competing population.

Factor, abiotic. A physical or chemical (hence nonliving) feature of the environment that influences the growth and survival of organisms.

   biotic. An environmental feature due to the presence of one or more other living organisms in the same habitat.

    density-dependent. An environmental feature (usually biotic) which becomes increasingly important with a rise in the species population.

    density-independent. An environmental feature (usually abiotic) that is equally effective when few or many other members of the same species are present.

    density-limiting. An environmental feature that sharply curtails the number of reproducing members of a species population in a limited region.

Factor interaction. Modification of the effect of one environmental factor by a change in another factor, such as the interaction between wind and humidity.

Family. A level in classification below an order and above a genus.

Feedback. An effect of the product of a process on the rate of the process, as a form of automatic regulation of rate.

Flyoff water. Precipitation that evaporates and is carried away by moving air before being absorbed by soil or any organisms.

Food web. Network of nutritive relationships among members of all ages and all species populations in a community.

Forb. Any nonwoody plant other than a grass.

Gall, plant. An abnormal growth on a vascular plant in response to an invading organism or some product from it.

Gallery forest. A strip of forest on a river or lake, where tree growth is supported by water flowing through the soil for a short distance.

Genus. A level of classification between a family and species.

Geophyte. A biennial or perennial vascular plant whose buds are located below the surface of the soil on subterranean stems.

Gleization. A combination of physical, chemical and biological changes in polar tundra soils, producing a sticky, compact, structureless layer ("glei").

Gnotobiology. The study of multicellular organisms grown in a microbe-free environment.

Green revolution. The spectacular increase in food production that is being sought to match the needs of present and future human populations.

Greenhouse effect. The retention of heat in and below the earth's atmosphere due to absorption of longwave radiations traveling toward outer space.

Hadal. Pertaining to the greatest depths of the oceans, in trenches more than 10,000 feet below the surface.

Halophyte. Any plant tolerant of high salinity in water or soil, and often of salt spray as well.

Hardpan. The calcareous deposit that forms a watertight stratum below the soil of many grasslands.

Hardwood. A tree of class Angiospermae, the flowering plants, regardless of the hardness of the wood.

Herbaceous plant. A vascular plant that dies down to the ground level at the approach of winter.

Herbivore. An animal that eats primarily foods of vegetable origin.

Heterotrophic organism. Any form of life that gets its energy and nutrient materials from the live or dead bodies of other organisms.

Homology, embryonic. A similarity in developmental origin, often of structures with ultimate dissimilarity in function and appearance.

Human ecology. The study of environmental factors affecting the welfare of mankind, as a special form of autecology.

Hydrologic cycle. The repeated movement of water by solar evaporation into dispersing winds, followed by precipitation and return to the sea.

Hydrophyte. A vascular plant that grows in water with its buds below the water surface.

Hyperparasite. An organism parasitic upon a host that is itself a parasite.

Hypolimnion. The cooler, darker and less aerated water in a lake below the mesolimnion (thermocline) and hence beyond the mixing effect of wave action.

Hypoxia. Reduced availability of oxygen, usually at high elevations where the partial pressure of the gas is too low to move it rapidly into animal bodies.

Immigration. The arrival of organisms foreign to a region in a dispersing movement past some natural barrier.

Inertia, thermal. A characteristic of water, which can absorb or release large amounts of heat without great change in temperature.

Isolation, genetic. Separation of a portion of a species population from the principle breeding stock, ending the reciprocal flow of genes.

Juvenile hormone. An endocrine secretion of immature insects, which prolongs the nutritive early stages of development and retards maturation.

Krummholz. The twisted, gnarled trees and shrubs along the boundary between forest and tundra in alpine environments.

Kwashiorkor. A distinctive pattern of abnormal growth which develops soon after weaning in children whose diet contains an inadequate amount or variety of proteins.

Laterization. The combination of rapid decomposition and leaching of soluble material in soils of tropical rain forests, leaving a reddish yellow residue (laterite).

Law of the inoptimum. No species encounters in any habitat the optimum conditions for all of its functions.

Law of the minimum. The growth of an organism is limited by whatever nutrient material is available to it in minimum quantity.

Law of the optimum. An organism finds its most suitable habitat where its tolerances are least under stress by the many factors of the environment.

Leaching action. The removal of soluble compounds from soil by cutoff water as it passes through into aquifers.

Leaf wash. The soluble material excreted or deposited on the bark, and transported to the soil by precipitation.

Lethal limit. An extreme environmental condition that causes death to organisms continually exposed to it.

Limnology. The study of ecological aspects of freshwater systems.

Littoral zone. The shallow coastal waters where life is affected by wave action, tidal changes, and runoff from the land.

Logistic curve. A graphical representation of change in populaton size, due first to unrestricted reproduction and then restriction by environmental factors.

Long-day plants. An angiosperm that shows a photoperiodic response by flowering when days are long and nights short.

Macronutrient. A dissolved inorganic substance required in greater than trace amounts for normal growth by a plant.

Matrix. Anything that provides spacing and continuity to a system, as a framework.

Mesolimnion. The intermediate stratum between epilimnion and hypolimnion, characterized by rapid change in temperature and oxygen concentration.

Mesophyte. A terrestrial plant that is adapted to live where soil water is moderately available.

Metamorphosis. A transformation in body form from the characteristics of larval stages to those of maturity.

Metabolic water. A product of chemical transformations in digestion that takes the place of water obtained by drinking.

Microbiota. All of the species populations of protozoans, bacteria and inconspicuous fungi in an area or geological period.

Microclimate. The combination of temperature, moisture supply and other environmental conditions in the space around an organism.

Microhabitat. That portion of the physical environment actually inhabited by the members of a species population, excluding adjacent areas that are not used.

Micronutrient. A dissolved substance required in only trace amounts for normal growth.

Migration. Repeated journeys by the same individual between separate areas at different parts of the year or of the life cycle.

Millimicron (see Nanometer)

Monoclimax. A self-perpetuating plant community in which the overall influence of the regional climate obscures any local variations due to the substrate.

Monoculture. The replacement of natural mixed communities of many species with a single dominant population of a cultigen which is to be harvested as a crop.

Mutualism. A social interaction between members of different species, benefitting both partners, but necessary for at least one partner.

Mycorrhiza. A fungus associating with the roots of a vascular plant, either surrounding the root tip or invading root cells in a mutualistic relationship.

Nanometer. A billionth of a meter or a millionth of a millimeter or a thousandth of a micron; hence formerly called a millimicron.

Natality. The characteristic number of births, hatchings, or germinations among members of a species population in a given time.

Nekton. Aquatic animals that swim in pursuit of specific prey or in directions other than vertical, rather than that drift passively with currents.

Neritic province. That portion of the marine environment that lies above the continental shelves, as contrasted to the province of the open ocean.

Neuston. Organisms (individually neusters) that spend major portions of their lives associated with the surface film over aquatic environments.

Neutralism. An association between two or more species populations in which no species benefits or is harmed by the presence of the others.

Niche, ecological. The totality of requirements in habitat and nutrition characteristic of members of a species population.

Nitrification. The progressive oxidation by soil microbes of compounds containing nitrogen, such as ammonia, to nitrates and nitrites.

Nitrogen-fixation. The incorporation of dissolved nitrogen into nitrogenous compounds, such as is known to occur in certain blue-green algae, bacteria and fungi.

Nodule, root. A characteristic enlargement (gall) on the root of a vascular plant within which nitrogen-fixing organisms live mutualistically.

Nutrient trap. A concentration of important nutrient materials, chiefly organic, on the bottom at the mouth of an estuary, due partly to the pattern of currents.

Oligotrophic lake. A body of fresh water in which dissolved materials, both inorganic and organic, are all at low concentrations.

Order. A level in classification below a class and above a family.

Osmoregulation. An adaptive mechanism that protects the balance between water and solutes within a cell or organism despite salinity changes in the environment.

Overturn, lake. A seasonal mixing of waters, obliterating stratification temporarily, when the temperature reaches that corresponding to the maximum density of water.

Ovoviviparous. Retaining the eggs until active young can be released, but with no nutritional contribution from parent to offspring after the eggs are fertilized.

Pedology. The study of soils as physico-chemical systems and as products of geological and biological activities.

Pelagic organism. Among the nekton any form of aquatic animal that lives remote from the shores and bottom.

Periodicity. A somewhat regular rise and fall in a species population, or an analogous change in activity of individuals in a population.

Permafrost. A layer of mineral particles embedded in ice that underlies polar soils as a condition that began in Pleistocene times.

Phanerophyte. Perennial woody plants with buds more

than 10 inches above the soil surface and with roots in the soil.

Phenology. A comparative study of the timing of seasonal events in plants, such as appearance of leaves on deciduous trees, of flowers, or ripe fruits.

Pheromone. An organic compound or mixture with volatile components serving in communication among members of the same species of animal.

Photoperiodism. A response in plant growth or animal behavior to daily, lunar or seasonal changes in illumination.

Phylum. A primary subdivision of the plant kingdom or animal kingdom, with each phylum composed of one or more classes.

Physiological ecology. The study of adaptive features and adjustments in the functions of organisms corresponding to changes in environmental conditions.

Phytoedaphon. Small or microscopic plants within the soil, often referred to as the "flora" of the soil although none is a flowering plant.

Phytoplankton. Small or microscopic plants that drift about in aquatic environments, chiefly in the illuminated euphotic zone.

Plankton. Small or microscopic organisms that drift about in aquatic environments.

Podzolization. A process of soil formation under trees with resinous needles, which resist decomposition but acidify the soil, expediting leaching.

Polyclimax. A self-perpetuating community of plants in which the species composition shows differences according to the succession that produced it.

Population ecology. The study of reactions in species populations to environmental conditions.

Potential, biotic. The totality of characteristics in a species that promote its reproduction, such as number of eggs or young in a litter and of litters in a lifetime.

Preadaptation. A functional or structural feature that arises prior to an adaptive use and later becomes subject to natural selection.

Precocial. Able to fend for itself—as of a chick that can detect and pick up its own food and respond behaviorally to many adverse changes in the environment.

Producer. An organism that synthesizes from inorganic materials a number of organic compounds containing recoverable energy; chiefly green plants engaged in photosynthesis.

Productivity, ecosystem. A quantitative measure of the rate of production of new biomass or of energy storage at any stage in an ecosystem.

   photosynthetic. A measure of the solar energy trapped by a green plant.

Profile, soil. A vertical section downward from the soil surface.

Programming, genetic. The sequential release at the molecular level in cells of biosynthetic and biodegrative processes that control growth and death.

Protist. A member of the Kingdom Protista, defined

variously to include some or all of the protozoans, bacteria and rickettsias, unicellular and colonial algae, inconspicuous fungi, and other thallophyte plants.

Protocooperation. An optional form of interaction between members of different species, benefitting both partners.

Pyramid, age. A graphic representation showing by the width and area of each band in a vertical tier the number of individuals in each age group of a population.

   biomass. A graphic representation showing by the width and area of each band in a vertical tier the total biomass in each trophic level of an ecosystem.

   energy. A graphic representation showing by the width and area of each band in a vertical tier the stored energy in each trophic level of an ecosystem.

   numbers. A graphic representation showing by the width and area of each band in a vertical tier the number of individuals in each trophic level of an ecosystem.

Radiation, adaptive. The divergence in body function and form among related organisms that have evolved these features in relation to their ecological niches.

Realm, biogeographic. A subdivision of continental scope made distinctive by its endemic orders, families and genera of living things.

Recycling. The natural or man-controlled reconstitution of chemical materials into a form that can be reused.

Renewable resource. Any form of life from the population of which a regular harvest can be taken repeatedly without endangering the breeding stock; sometimes used also of soil and other products of living communities.

Resistance, environmental. The totality of ecological interactions that ordinarily limit any continued rise in a species population to the carrying capacity.

Rhizome. An underground stem, usually horizontal, in which food reserves are stored and from which branches or foliage grow upward into air.

Rhythm, circadian. A periodic change in function or behavior of animals that continues for a while on an almost daily cycle under constant conditions.

   tidal. A periodic change in function or behavior that continues on an almost tidal (23-hours-plus-9-minutes) cycle under constant conditions in confinement.

Rule, Allen's. The extremities (beak, ears, legs, and tail) of warm-blooded animals are usually shorter in cool parts of a continuous range than in warm parts.

   Bergmann's. The average size of adult warm-blooded animals is larger in cool parts of a continuous range than in warm parts.

   Jordan's. The number of vertebrae in fishes with a wide north–south range is greater where spawning is in cool water than where it occurs in warm water.

Runoff water. Water in the hydrologic cycle that returns to the ocean from the land by way of exposed rivers and lakes.

Salination, soil. A process characteristic of most desert soils in which evaporation from the surface tends to leave a saline crust in the uppermost levels.

Salinity. The concentration of ions, primarily of sodium, magnesium, chloride, and sulfate, in sea water.

Saprophyte. A unicellular or multicellular plant that obtains its nourishment from dead organisms or organic residues; hence a decomposer.

Selection, natural. The elimination of individuals in a population under stress in proportion to their intolerance for the particular stress.

Seral stage. A characteristic association of plants and animals during an ecological succession after colonization and before climax is reached.

Short-day plant. An angiosperm that shows a photoperiodic response by flowering when days are short and nights long.

Softwood. A term applied in the lumber industry to any coniferous tree regardless of the softness or hardness of its wood.

Speciation. The evolution of species through genetic isolation, establishment of a distinctive ecological niche, and natural selection of adaptive features.

Species. A "kind" of organism, implying interfertility in sexual reproduction, geographic continuity in the recent past or present, and common ancestry.

Spring tide. The particularly large tidal changes that occur when the gravitational forces of moon and sun are combined, as at new moon and full moon.

Stenohaline. Tolerance of only a narrow range of change in salinity of the environment, without specifying whether this is at high salinity or low.

Stenothermal. Tolerance of only a narrow range of change in temperature, without specifying where this range of temperature falls on the scale.

Stratification. The aggregation of members of particular species into one horizontal layer in a vertical sequence in an aquatic environment, the soil, or among terrestrial vegetation.

Stress, physiological. A challenge to the inherited tolerances of an organism to environmental changes, such as in temperature, water supply or salinity.

Succession, ecological. The progressive change in the species composing the ecological community from colonization through seral stages to climax.

   primary. Successional changes following exposure of new land area, as by landslide or glacier retreat or lava flow, where no organic matter is present.

   secondary. Successional changes following abandonment of a farm or other area in which soil is present, with humus and soil organisms.

Succulence. Unusual development of water storage tissues by many plants in arid or physiologically dry environments.

Swamp. Any wetland dominated by woody shrubs and trees.

Symbiosis. An association of two unrelated species, without indication that members of one species benefit or are harmed by the association.

Sympartic species. Plants or animals whose geographic

ranges overlap, a term employed in connection with potential hybridization of close relatives.

Synecology. The study of interrelationships among all kinds of organisms in an ecosystem in relation to the environment.

Synthetic mode. A rational attempt to combine into an understandable scheme all of the available information on a subject.

Systems ecology. The study of interrelationships among living things and their abiotic environment in operational terms with binary alternatives.

Taiga. The northern (boreal) coniferous forest bordered on the north by tundra.

Territoriality. The behavior of animals that defend against intruders of their own species a space around their eggs or young.

Thermocline, or mesolimnion. The stratum in a deep pond or lake where the temperature and oxygen concentration diminish rapidly.

Therophyte. An annual plant in which the only cells surviving winter or summer drought are those in seeds or spores.

Trophic level. A stratum in the ecological hierarchy of producers, consumers, and decomposers, as shown graphically in a pyramid of biomass or energy.

Tundra. An area at high latitude or high altitude where climatic conditions are too severe for upright growth of woody plants.

Turnover rate or time. A measure of lifespan in terms of the interval between absorption of mineral nutrients by producers and their release by decay.

Vernalization. Exposure to a period of prolonged cold as a normal way to release in certain seeds the response to moisture shown in germination.

Virus. An infective particle consisting of nucleic acid (DNA or RNA) and protein, capable of inducing a susceptible host cell to synthesize new viruses.

Waste. Disseminules that contribute nothing because they do not survive, and food that is eaten without being digested and absorbed.

Work loop. The amount of energy that an organism must accumulate before reaching the reproductive stage.

Xerophyte. A plant that can live and reproduce with a minimum of water, through adaptive features such as succulence.

Year-class. All individuals originating in the same annual reproductive season, as a cohort whose mortality can be followed separately.

Yield, perpetual. A program of planting and harvesting designed to allow production of a crop each year over an indefinite period with no loss of quality.

Zoochlorella. Algal cells, mutualistic in sponges, corals, or other multicellular animals.

Zoonosis. A disease of wild animals that can be contracted by man or domesticated animals because the parasitic agents can use either host.

Zooplankton. The animals among the drifting life (plankton) in upper levels of an aquatic environment.

Zooxanthella. Brownish or yellowish unicellular plants, chiefly dinoflagellates, mutualistic in marine animals.

Bibliography

Chapter 1. Beginnings of Ecology

Kormondy, E. J., *Readings in Ecology* (Englewood Cliffs: Prentice-Hall, 1965)

Chapter 2. Energy for Life

Cadle, R. D., and E. R. Allen, "Atmospheric photochemistry," *Science* (Jan. 16, 1970)

Gates, David M., "Spectral distribution of solar radiation at the earth's surface," *Science* (Feb. 4, 1966)

Goldman, Charles R., "Aquatic primary production," *American Zoologist* (Feb. 1968)

Woodwell, G. M., and R. H. Whittaker, "Primary production in terrestrial ecosystems," *American Zoologist* (Feb. 1968)

Chapter 3. Consumers

Conover, Robert J., "Zooplankton—life in a nutritionally dilute environment," *American Zoologist* (Feb. 1968)

Darnell, Rezneat M., "Animal nutrition in relation to secondary production," *American Zoologist* (Feb. 1968)

Engelmann, Manfred D., "The role of soil arthropods in community energetics," *American Zoologist* (Feb. 1968)

Gates, David M., "The energy environment in which we live," *American Scientist* (Sept. 1963)

Golley, Frank B., "Secondary productivity in terrestrial communities," *American Zoologist* (Feb. 1968)

Read, Clark P., "Some aspects of nutrition in parasites," *American Zoologist* (Feb. 1968)

Slobodkin, Lawrence B., "How to be a predator," *American Zoologist* (Feb. 1968)

Chapter 4. Cycles, Pyramids, Niches

Bolin, Bert, "The carbon cycle," *Scientific American* (Sept. 1970)

Bormann, F. H., and G. E. Likens, "The nutrient cycles of an ecosystem," *Scientific American* (Oct. 1970)

Cloud, Preston, and Aharon Gibor, "The oxygen cycle," *Scientific American* (Sept. 1970)

Cole, LaMont C., "The ecosphere," *Scientific American* (April 1958)

Deevey, Edward S. Jr., "Mineral cycles," *Scientific American* (Sept. 1970)

Delwiche, C. C, "The nitrogen cycle," *Scientific American* (Sept. 1970)

Hutchinson, G. Evelyn, "The biosphere," *Scientific American* (Sept. 1970)

Likens, G. E., *et al.*, "The calcium, magnesium, potassium, and sodium budgets for a small forested ecosystem," *Ecology* (Summer, 1967)

Milne, Lorus J., and Margery Milne, *Water and Life* (New York: Atheneum, 1964)

Odum, Eugene P., "Energy flow in ecosystems," *American Zoologist* (Feb. 1968)

Odum, Howard T., "Ecological potential and analogue circuits for the ecosystem," *American Scientist* (March 1960)

Penman, H. L., "The water cycle," *Scientific American* (Sept. 1970)

Chapter 5. The Physical Challenges to Life

Gates, David M., "Heat transfer in plants," *Scientific American* (Dec. 1965)

Hendricks, Sterling B., "Metabolic control of timing," *Science* (July 5, 1963)

Hendricks, Sterling B., "How light interacts with living matter," *Scientific American* (Sept. 1968)

Menaker, Michael, "Biological clocks," *BioScience* (Aug. 1969)

Moen, Aaron N., "The critical thermal environment," *BioScience* (Nov. 1968)

Wald, George, "Life and light," *Scientific American* (Oct. 1959)

Chapter 6. Avoiding Conflict Within the Species

Allee, W C., *Cooperation among Animals, with Human Implications* (New York: Henry Schuman, 1951)

Ardrey, Robert, *The Territorial Imperative* (New York: Atheneum, 1966)

Durham, J. Wyatt, "The incompleteness of our knowledge of the fossil record," *Journal of Paleontology* (May 1967)

Lack, David, *Darwin's Finches: An Essay on the General Biological Theory of Evolution* (Cambridge: Cambridge University Press, 1947)

Mayr, Ernst, *Principles of Systematic Zoology* (New York: McGraw-Hill, 1969)

Newell, Norman D., "Adequacy of the fossil record," *Journal of Paleontology* (March 1959)

Simpson, George G., "How many species?" *Evolution* (Sept. 1952)

Chapter 7. How Different Species Live Together

Arditti, Joseph, "Orchids," *Scientific American* (Jan. 1966)

Blest, A. D., "Longevity, palatability and natural selection in five species of New World saturniid moth," *Nature* (March 23, 1963)

Brower, L. P., "Ecological chemistry," *Scientific American* (Feb. 1969)

Daubin, William H., "The tuatara in its natural habitat," *Endeavour* (Jan. 1962)

Dodson, Calaway H., *et al.*, "Biological active compounds in orchid fragrances," *Science* (June 13, 1969)

Erhlich, Paul R., and Peter H. Raven, "Butterflies and plants," *Scientific American* (June 1967)

Grant, Verne, "The fertilization of flowers," *Scientific American* (June 1951)

Kettlewell, H. B. D., "Insect survival and selection for pattern," *Science* (June 4, 1965)

Kudairi, A. K., "Mycorrhiza in desert soils," *BioScience* (July 1969)

Lamb, I. Mackenzie, "Lichens," *Scientific American* (Oct. 1959)

Limbaugh, Conrad, "Cleaning symbiosis," *Scientific American* (Aug. 1961)

Went, F. W., and N. Stark, "Mycorrhiza," *BioScience* (Nov. 1968)

Wilde, S. A., "Mycorrhizae and tree nutrition," *BioScience* (June 1968)

Chapter 8. Growth of Populations

Andrewartha, H. G., and L. C. Birch, *The Distribution and Abundance of Animals* (Chicago: University of Chicago Press, 1954)

Comfort, Alex, "The life span of animals," *Scientific American* (Aug. 1961)

Comfort, Alex, *Ageing: The Biology of Senescence* (New York: Holt, Rinehart & Winston, 1964)

Desmond, Annabelle, "How many people have ever lived on Earth?" *Population Bulletin* (1962)

Dice, Lee R., *Natural Communities* (Ann Abor: University of Michigan Press, 1952)

Djerassi, Carl, "Birth control after 1984," *Science* (Sept. 4, 1970)

Hardin, Garrett, "The competitive exclusion principle," *Science* (April 29, 1960); criticisms. July 8, 1960, Aug. 5, 1960, Dec. 2, 1960, Dec. 9, 1960

Hazen, William E. (ed.) *Readings in Population and Community Ecology* (Philadelphia: W. B. Saunders, 1964)

Lack, David, *The Natural Regulation of Animal Numbers* (Oxford: Clarendon Press, 1954)

MacArthur, Robert H., and Joseph H. Connell, *The Biology of Populations* (New York: John Wiley & Sons, 1966)

Milne, Lorus J., and Margery Milne, *The Ages of Life: A New Look at the Effects of Time on Mankind and Other Living Things* (New York: Harcourt, Brace, and World, 1968)

Pimentel, David, "Population regulation and genetic feedback," *Science* (March 29, 1968)

Sacher, G. A., "Relation of lifespan to brain weight and body weight in mammals," in *The Life Span of Animals* (Boston: Little, Brown & Co., 1959)

Slobodkin, Lawrence B., *Growth and Regulation of Animal Populations* (New York: Holt, Rinehart & Winston, 1961)

Wynne-Edwards, V. C., "Population control in animals," *Scientific American* (Aug. 1964)

Wynne-Edwards, V. C., "Self-regulating system in populations of animals," *Science* (March 26, 1965); criticisms May 14, 1965.

Chapter 9. The Human Environment

Altman, P. L., and D. S. Dittmer (eds.), *Growth: Including Reproduction and Morphological Development* (Washington, D. C.: Federation of American Societies for Experimental Biology, 1962)

Boerma, Addeke H., "A world agricultural plan," *Scientific American* (Aug. 1970)

Bresler, Jack B. (ed.) *Human Ecology: Collected Readings* (Reading: Addison-Wesley, 1966)

Bresler, Jack B. (ed.) *Environments of Man* (Reading: Addison-Wesley, 1968)

Brown, Harrison, *The Challenge of Man's Future* (New York: Viking, 1954)

Brown, Harrison, "Human materials production as a process in the biosphere," *Scientific American* (Sept. 1970)

Brown, Lester R., "Human food production as a process in the biosphere," *Scientific American* (Sept. 1970)

Burkill, I. H., "Habits of man and the origins of the cultivated plants of the Old World," *Proceedings of the Linnean Society of London* (1953)

Caspari, Ernst, "Selective forces in the evolution of man," *American Naturalist* (1963)

Cole, LaMont C., "Man's ecosystem," *BioScience* (April 1966)

Coon, Carleton, "Climate and race," *Smithsonian Annual Report* for 1953

Crowe, Beryl L., "The tragedy of the commons revisited," *Science* (Nov. 28, 1969)

Daniels, Farrington, "Direct use of the sun's energy," *American Scientist* (Spring, 1967)

Darling, F. Fraser, and John P. Milton (eds.), *Future Environments of North America* (Garden City: Natural History Press, 1966)

Dasmann, Raymond F., *Environmental Conservation* (New York: John Wiley & Sons, 1968)

Ehrlich, Paul R., *The Population Bomb* (New York: Ballantine's, 1968)

Ehrlich, Paul R., and Anne H. Ehrlich, *Population: Resources: Environment: Issues in Human Ecology* (San Francisco: W. F. Freeman, 1970)

Ehrlich, Paul R., and John P. Holdren, "Population and panaceas: a technological perspective," *BioScience* (Dec. 1969)

Fertig, Daniel S., and Vaughan W. Edmonds, "The physiology of the house mouse," *Scientific American* (Oct. 1969)

Gregg, Alan, "Is man a biological cancer?" *Population Bulletin* (1955)

Handler, Philip (ed.) *Biology and the Future of Man* (New York: Oxford University Press, 1970)

Hardin, Garrett, *Nature and Man's Fate* (New York: Henry Holt, 1959)

Hardin, Garrett, "The tragedy of the commons," *Science* (Dec. 13, 1968)

Hardin, Garrett (ed.) *Population, Evolution, and Birth Control* (San Francisco: W. H. Freeman, 1969)

Howells, William W., "The distribution of man," *Scientific American* (Sept. 1960)

Istock, Conrad. "A corollary to the dismal theorem," *BioScience* (Dec. 1969)

Jordan, P. A., "Ecology, conservation, and human behavior," *BioScience* (Nov. 1968)

Krutch, Joseph Wood, "Conservation is not enough," *American Scholar* (Summer, 1954)

Lasker, Gabriel W., "Human biological adaptability," *Science* (Dec. 19, 1969)

Martin, Paul S., "Pleistocene niches for alien animals," *BioScience* (Feb. 15, 1970)

Milne, Lorus J., and Margery Milne, *Water and Life* (see Chap. 4)

Milne, Lorus J., and Margery Milne, *The Ages of Life* (see Chap. 8)

Moncrief, L. W., "The cultural basis of our environmental crisis," *Science* (Oct. 30, 1970)

Paddock, William C., "How green is the Green Revolution?" *BioScience* (Aug. 15, 1970)

Paddock, William, and Paul Paddock, *Hungry Nations* (Boston: Little, Brown, 1964)

Paddock, William, and Paul Paddock, *Famine—1975! America's Decision* (Boston: Little, Brown, 1967)

Pinchot, G. B., "Marine farming," *Scientific American* (Dec. 1970)

Reed, Charles A., "Animal domestication in the prehistoric Near East," *Science* (Dec. 11, 1959)

Revelle, Roger, and Hans H. Landsberg (eds.) *America's Changing Environment* (Boston: Houghton Mifflin, 1970)

Rosen, W. G., "The environmental crisis," *BioScience* (Nov. 15, 1970)

Schreider, Eugene, "Ecological rules, body heat regulation, and human evolution," *Evolution* (Jan. 1964)

Sharman, G. B., "Reproductive physiology of marsupials," *Science* (Feb. 27, 1970)

Shepard, Paul, and Daniel McKinley (eds.) *The Subversive Science: Essays toward an Ecology of Man* (New York: Houghton Mifflin, 1969)

Simpson, George Gaylord, "The nonprevalence of humanoids," *Science* (Feb. 21, 1964)

Singer, S. Fred, "Human energy production as a process in the biosphere," *Scientific American* (Sept. 1970)

Slobodkin, Lawrence B., "Aspects of the future of ecology," *BioScience* (Jan. 1968)

Spengler, Joseph J., "Population problem: in search of a solution," *Science* (Dec. 5, 1969)

Taylor, C. R., "The eland and the oryx," *Scientific American* (Jan. 1969); comments (Nov. 1969)

Thomas, William L., Jr. (ed.) *Man's Role in Changing the Face of the Earth* (Chicago: University of Chicago Press, 1956)

Ucko, Peter J., and G. W. Dimbleby (eds.), *The Domestication and Exploitation of Plants and Animals* (London: Aldine, 1969)

Waterbolk, H. T., "Food production in prehistoric Europe," *Science* (Dec. 6, 1968)

White, Lynn, Jr., "The historical roots of our ecologic crisis," *Science* (March 10, 1967); comments (May 12, 1967)

Wright, H. E., Jr., "Natural environment of early food production north of Mesopotamia," *Science* (July 26, 1968).

Yarwood, C. E., "Man-made plant diseases," *Science* (April 10, 1970)

Young, Gale, "Dry lands and desalted water," *Science* (Jan. 23, 1970)

Chapter 10. Ecological Awareness

Allen, A. A., R. S. Schlueter, and P. G. Mikolaj, "Natural oil seepage at Coal Oil Point, Santa Barbara, California," *Science* (Nov. 27, 1970)

Carson, Rachel, *Silent Spring* (Boston: Houghton Mifflin, 1962)

Cole, LaMont C., "Thermal pollution," *BioScience* (Nov. 1969); comments (Jan. 15, 1970)

Cornwall, I. W., *Prehistoric Animals and Their Hunters* (New York: Praeger, 1968)

Cox, J. L., "DDT residues in marine phytoplankton: increase from 1955 to 1969," *Science* (Oct. 2, 1970)

Graham, Frank, Jr., *Since Silent Spring* (Boston: Houghton Mifflin, 1970)

Harrison, H. L., *et al.*, "Systems studies of DDT transport," *Science* (Oct. 30, 1970)

Harriss, R. C., D. B. White, and R. B. Macfarlane, "Mercury compounds reduce photosynthesis by plankton," *Science* (Nov. 13, 1970)

Krantz, Grover S., "Human activities and megafaunal extinctions," *American Scientist* (March/April 1970)

Lave, Lester B., and Eugene P. Seskin, "Air pollution and human health," *Science* (Aug. 21, 1970)

McNulty, Faith, "The prairie dog and the black-footed ferret," *New Yorker* (June 13, 1970)

Murray, E. G. D., "The place of nature in man's world," *American Scientist* (Jan. 1954)

Pruitt, William O., Jr., "Lichen, caribou, and high radiation in Eskimos," *Audubon Magazine* (Sept./Oct. 1963)

Woodwell, George M., "Toxic substances and ecological cycles," *Scientific American* (March 1967)

Ziswiler, Vinzenz, *Extinct and Vanishing Animals* (New York: Springer-Verlag, 1967)

Chapter 11. Interactions within the Environment

Elton, Charles S., *The Ecology of Invasions by Animals and Plants* (New York: John Wiley & Sons, 1958)

Powers, Charles F., and Andrew Robertson, "The aging Great Lakes," *Scientific American* (Nov. 1966)

Chapter 12. The Marine World

Baker, D. James, Jr., "Models of oceanic circulation," *Scientific American* (Jan. 1970)

Bates, Marston, *The Forest and the Sea* (New York: Random House, 1960)

Berger, Wolfgang H., "Foraminiferal ooze: solution at depths," *Science* (April 21, 1967)

Berger, Wolfgang H., "Planktonic foraminifera: field experiment on production rate," *Science* (June 16, 1967)

Berger, Wolfgang H., "Radiolarian skeletons: solution at depths," *Science* (March 15, 1968)

Bullard, Sir Edward, "The origin of the oceans," *Scientific American* (Sept. 1969)

David, P. M., "The surface fauna of the ocean," *Endeavour* (May 1965)

Dietz, Robert S., "The sea's deep scattering layers," *Scientific American* (Aug. 1962)

Emery, K. O., "The continental shelves," *Scientific American* (Sept. 1969)

Fraser, James, *Nature Adrift: The Story of Marine Plankton* (Philadelphia: Dufour Editions, 1962)

Hardy, Sir Alistair, *The Open Sea, its Natural History: Part 1. The World of Plankton* (London: Collins, 1956)

Hardy, Sir Alistair, *The Open Sea, its Natural History, Part 2. Fish and Fisheries* (London: Collins, 1960)

Holt, S. J., "The food resources of the ocean," *Scientific American* (Sept. 1969)

Isaacs, John D., "The nature of oceanic life," *Scientific American* (Sept. 1969)

MacIntyre, Ferren, "Why the sea is salt," *Scientific American* (Nov. 1970)

Marshall, N. B., *Aspects of Deep-Sea Biology* (New York: Philosophical Library, 1954)

Menard, H. W., "The deep-ocean floor," *Scientific American* (Sept. 1969)

Redfield, Alfred C., "Ontogeny of a salt marsh estuary," *Science* (Jan. 1, 1965)

Ryther, John H., "Photosynthesis and fish production in the sea," *Science* (Oct. 3, 1969); comments (April 14, 1970)

Sanders, Howard L., and Robert R. Hessler, "Ecology of the deep-sea benthos," *Science* (March 28, 1969); criticisms (Nov. 21, 1969)

Schmidt-Nielsen, Knut, "Salt glands," *Scientific American* (Jan. 1959)

Sillén, Lars G., "The ocean as a chemical system," *Science* (June 2, 1967)

Stephens, Grover C., "Dissolved organic matter as a potential source of nutrition for marine organisms," *American Zoologist* (Feb. 1968)

Stewart, R. W., "The atmosphere and the ocean," *Scientific American* (Sept. 1969)

Sverdrup, H. V., M. W. Johnson, and R. H. Fleming, *The Oceans: Their Physics, Chemistry, and General Biology* (Englewood Cliffs: Prentice-Hall, 1942)

Wolff, Torben, "Life in the ocean six miles down," *New Scientist* (Oct. 22, 1964)

Chapter 13. Life in Fresh Waters

Coker, Robert E., *Streams, Lakes and Ponds.* (Chapel Hill: University of North Carolina Press, 1954)

Deevey, Edward S., Jr., "Bogs," *Scientific American* (Oct. 1958)

Green, James, "Freshwater jellyfish," *New Scientist* (July 11, 1963)

Hutchinson, G. Evelyn, *A Treatise on Limnology. I. Geography, Physics, and Chemistry* (New York: John Wiley & Sons, 1966)

Hutchinson, G. Evelyn, *A Treatise on Limnology. II. Introduction to Lake Biology: the Plankton* (New York: John Wiley & Sons, 1966)

Palmer, E. Laurence, "Stream edges," *Natural History* (April 1961)

Welch, Paul S., *Limnology* (New York: McGraw-Hill, 1962)

Chapter 14. The Hidden Life of the Soil

Borman, F. H., and G. E. Likens, "Nutrient cycling," *Science* (Jan. 27, 1967)

Farb, Peter, *Living Earth* (New York: Harper, 1959)

Poulson, Thomas L., and William B. White, "The cave environment," *Science* (Sept. 5, 1969)

Savory, Theodore H., "Hidden lives," *Scientific American* (July 1968)

Stewart, W. D. P., "Nitrogen-fixing plants," *Science* (Dec. 15, 1967)

Went, F. W., "Fungi associated with stalactite growth," *Science* (Oct. 17, 1969)

Chapter 15. Life in the Forests

Bates, Marston, *The Forest and the Sea* (New York: Random House, 1960)

Cooper, Charles F., "The ecology of fire," *Scientific American* (April 1961)

Dobzhansky, Theodosius, and João Murca-Pires, "Strangler trees," *Scientific American* (Jan. 1954)

Golley, F. B., *et al.*, "The structure of tropical forests in Panama and Colombia," *BioScience* (Aug. 1969)

McCormick, Jack, *The Life of the Forest* (New York: McGraw-Hill, 1966)

Chapter 16. Grassland Communities

Allen, Durward, *The Life of the Prairies and Plains* (New York: McGraw-Hill, 1967)

Chapter 17. The Living Deserts

Bartholomew, George A., and Jack W. Hudson, "Desert ground squirrels," *Scientific American* (Nov. 1961)

Bentley, P. J., "Adaptations of amphibia to arid environments," *Science* (April 29, 1966)

James, W. O., "Succulent plants," *Endeavour* (April 1958)

Niering, W. A., R. H. Whittaker, and C. H. Lowe, "The saguaro: a population in relation to environment," *Science* (Oct. 4, 1963)

Schmidt-Nielsen, Knut, *Desert Animals: Physiological Problems of Heat and Water* (New York: Oxford University Press, 1964)

Went, Frits W., "The ecology of desert plants," *Scientific American* (April 1955)

Chapter 18. Life at its Bleakest

Alexander, Gordon, and John R. Hilliard, Jr., "Altitudinal and seasonal distribution of Orthoptera in the Rocky Mountains of northern Colorado," *Ecological Monographs* (Autumn, 1969)

Baker, Paul T., "Human adaptation to high altitude," *Science* (March 14, 1969)

Corbet, Philip S., "Terrestrial microclimate: amelioration at high latitudes," *Science* (Nov. 14, 1969)

Dunbar, M. J., *Ecological Development in Polar Regions* (Englewood Cliffs: Prentice-Hall, 1968)

Irving, Laurence, "Adaptations to cold," *Scientific American* (Jan. 1966)

Llano, George A., "The terrestrial life of the Antarctic," *Scientific American* (Sept. 1962)

Murphy, Robert Cushman, "The oceanic life of the Antarctic," *Scientific American* (Sept. 1962)

Pruitt, William O., Jr., "Animals in the snow," *Scientific American* (Jan. 1960)

Ruud, Johan T., "The ice fish," *Scientific American* (Nov. 1965)

Sjörs, H., "Surface patterns in Boreal peatland," *Endeavour* (Oct. 1961)

Swan, Lawrence W., "The ecology of the high Himalayas," *Scientific American* (Oct. 1961)

Chapter 19. The Present Addresses of Living Things

Bates, Marston, "Man as an agent in the spread of organisms," *Man's Role in Changing the Face of the Earth*, W. L. Thomas, Jr. (ed.) (Chicago: University of Chicago Press, 1956)

Carlquist, Sherwin, *Island Life: A Natural History of the Islands of the World* (Garden City: Natural History Press, 1965)

Dansereau, Pierre, *Biogeography: An Ecological Perspective* (New York: Ronald Press, 1957)

Darlington, Philip J., Jr., *Zoogeography: The Geographical Distribution of Animals* (New York: John Wiley & Sons, 1957)

Darlington, Philip J., Jr., *Biogeography of the Southern End of the World* (Cambridge: Harvard University Press, 1965)

Dietz, R. S., and J. C. Holdren, "The breakup of Pangaea," *Scientific American* (Oct. 1970)

Dillon, Lawrence S., "The life cycle of the species: an extension of current concepts," *Systematic Zoology* (June 1966)

Engel, A. E. J., "Time and the earth," *American Scientist* (Winter, 1969)

Fischer, Alfred G., "Latitudinal variations in organic diversity," *Evolution* (March 1960)

Fleming, C. A., "Paleontology and southern biogeography," *Pacific Basin Biogeography—A Symposium*, J. Lindsley Gressitt (ed.) (Honolulu: Bishop Museum Press, 1963)

Good, Ronald, *The Geography of the Flowering Plants* (New York: John Wiley & Sons, 1964)

Haffer, Jürgen, "Speciation in Amazonian forest birds," *Science* (July 11, 1969)

Heirtzler, J. R., "Sea-floor spreading," *Scientific American* (Dec. 1968)

Hesse, Richard, W. C. Allee, and Karl P. Schmidt, *Ecological Animal Geography* (New York: John Wiley & Sons, 1951)

Hopkins, David M. (ed.) *The Bering Land Bridge* (Palo Alto: Stanford University Press, 1967)

Hurley, Patrick M., "The confirmation of continental drift," *Scientific American* (April 1968)

Hurst, G. W., "Meteorological aspects of insect migrations," *Endeavour* (May 1969)

Kurtén, Bjorn, "Continental drift and evolution," *Scientific American* (March 1969)

Lindroth, Carl H., *The Faunal Connections between Europe and North America* (New York: John Wiley & Sons, 1957)

Lindroth, Carl H., "Survival of animals and plants on ice-free refugia during the Pleistocene glaciations," *Endeavour* (Sept. 1970)

Mayr, Ernst, *Animal Species and Evolution* (Cambridge: Harvard University Press, 1963)

Mayr, Ernst, "Avifauna: turnover on islands," *Science* (Dec. 17, 1965)

Milne, Lorus J., and Margery Milne, *The Nature of Life* (New York: Crown, 1970)

Neill, Wilfred T., *The Geography of Life* (New York: Columbia University Press, 1969)

Newell, Norman D., "Crises in the history of life," *Scientific American* (Feb. 1963)

Simpson, G. G., *The Geography of Evolution* (Philadelphia: Chilton Books, 1965)

Walter, H., and E. Stadelmann, "The physiological prerequisites for the transition of autotrophic plants from water to terrestrial life," *BioScience* (July 1968)

Wells, Philip V., "Postglacial vegetational history of the great plains," *Science* (March 20, 1970)

Welty, Carl, "The geography of birds," *Scientific American* (July 1957)

Index

(Numbers in italics indicate photograph)

Prepared and produced by Chanticleer Press. *Publisher:* Paul Steiner. *Editor:* Milton Rugoff. *Associate Editor:* Jean Walker. *Art Director:* Ulrich Ruchti, assisted by Elaine Jones. *Production:* Gudrun Buettner, assisted by Ruth Charnes and Ursula Amrain. Printed by Amilcare Pizzi, S.p.A., Milan, Italy.

PHOTOGRAPHIC CREDITS

(read from left to right and from top to bottom)

Half-title page, Otto Croy; Title page montage: Arthur Tress, Andreas Feininger; Pages 9, 10, 12, 13, 14, American Museum of Natural History; 18, Rondal Partridge; 19, J. Allan Cash; 20, S. Jonansson; 21, Sacramento Peak Observatory; 22, Keith Gillett; 23, E. Aubert de La Rue, Pierre Pittet; 25, Emil Javorsky; 28, Emil Javorsky; 29, Emil Javorsky, Alexander Klots, Grant Heilman, H. Marx from Field Museum, Pierre Pittet, Herman Schunemann; 30, Stackhouse; 31, Grant Heilman, Eric Hosking; 35, Grant Heilman, Hugh Spencer; 38, Emil Javorsky; 39, Pierre Pittet, Emil Javorsky; 48, Jacana, Otto Croy; 49, Otto Croy, Willis Peterson, Lorus and Margery Milne, Rolf Blomberg from Full Hand, Lamont Geological Observatory; 52, E. Aubert de La Rue, W. T. Miller; 55, Arabian American Oil Company; 56, Tad Nichols; 57, Douglass Baglin Photography Pty Limited; 59, Duane D. Davis (both); 60, Alfred M. Bailey, T. Iwago; 62, Leonard Lee Rue III from National Audubon Society; 63, Lewis Walker; 64, Rondal Partridge, Leonard Lee Rue III, Edward S. Ross, Leonard Lee Rue III; 67, Jacana; 68, Otto Croy, Robert C. Hermes from National Audubon Society; 71, Australian News and Information Bureau, A. Schuhmacher, A. W. Ambler from National Audubon Society, Australian News and Information Bureau; 74, Emil Javorsky (both); 75, Edward S. Ross; 76, Leonard Lee Rue III, W. T. Miller; 77, Ed Park, Fritz Siedel; 80, Ivan Polunin, Leonard Lee Rue III; 81, John H. Gerard from National Audubon Society, Leonard Lee Rue III; 83, Dr. I. Eibl-Eibesfeldt; 86, Jack Dermid, G. Ronald Austing from National Audubon Society; 87, Popperfoto; 89, Redwood Empire Association; 90, Leonard Lee Rue III (both); 92, Eric Hosking, J. Allan Cash; 93, W. T. Miller, Leonard Lee Rue III, Wilhelm Hoppe; 96, J. Allan Cash; 97, Arthur Christiansen; 98, Alexander Klots; 99, Fritz Siedel, Hugh Spencer; 100, Leonard Lee Rue III, Allan D. Cruickshank from National Audubon Society; 102, Popperfoto; 103, Arthur Christiansen; 104, Grant Heilman, Leonard Lee Rue III; 105, Jacques Herissé, W. T. Miller; 106, Hugh Spencer (both); 107, Hugh Spencer (both); 112, Leonard Lee Rue III; 114, Kurt Severin; 115, Popperfoto (both); 116, Otto Croy, Grant Heilman, Lynwood Chace from National Audubon Society, Louis Quitt from National Audubon Society; 117, William E. Ferguson; 122, Popperfoto; 123, Allan D. Cruickshank from National Audubon Society; 125, Pierre Pittet; 126, Inger McCabe from Rapho Guillumette, Rolf Blomberg from Full Hand; 128, 129, Inger McCabe from Rapho Guillumette; 130, Arthur Tress; 137, Spence Air Photos; 140, Charles Herbert from Western Ways, Netherlands Information Service; 141, Pierre Pittet; 146, Inger McCabe from Rapho Guillumette; 150, U.S. Forest Service (both); 153, Marcia Keegan; 154, Alfred M. Bailey, Marcia Keegan, Marcia Keegan, Kurt Severin; 156, Rondal Partridge; 164, George Komorowski from National Audubon Society; 165, Eric Hosking from National Audubon Society, Douglass Baglin Photography Pty Limited; 166, John Hendry, Jr. from National Audubon Society; 169, Richard F. Conrat from Photophile; 170, Grant Heilman; 171, Inger McCabe from Rapho Guillumette; 172, Marcia Keegan; 173, Arthur Tress; 175, Richard F. Conrat from Photophile; 180, Arthur Christiansen; 181, Fritz Goro; 182, Leonard Lee Rue III; 183, Leonard Lee Rue III (both); 184, Lorus and Margery Milne; 186, Grant Heilman; 187, Jack Dermid, Gordon Smith; 188, WeHa-Photo Bern; 193, Philip Hyde; 197, Robert Ames from Photophile; 199, Grant Heilman (both); 200, Hal Harrison, Leonard Lee Rue III; 201, Otto Croy; 203, Charles E. Lane; 204, Popperfoto, Ron Church; 206, A. Schuhmacher, Marine Studios, Bernard Villaret; 209, Gulfarium, Florida; 210, Official Photograph U.S. Navy; 211, Woods Hole Oceanographic Institution, Fritz Siedel; 213, Douglas P. Wilson (all); 219, Jack Dermid; 220, René-Pierre Bille, Fritz Siedel; 221, Arthur Christiansen, Lorus and Margery Milne; 222, W. Schacht; 223, Lorus and Margery Milne, Ann and Myron Sutton; 224, National Park Service, Arthur Christiansen, Wolf Suschitzky; 231, Hal Harrison; 232, Karl H. Maslowski from National Audubon Society, Hugh Spencer, Henri Goldstein; 235, John H. Gerard from National Audubon Society, Alvin Staffer from National Audubon Society; 239, Emil Javorsky, Leonard Lee Rue III; 242, Ann and Myron Sutton, Grant Heilman; 243, E. Aubert de La Rue; 245, Ann and Myron Sutton; 246, Lorus and Margery Milne; 249, Philip Hyde; 250, Grant Heilman, Arthur Christiansen, J. Allan Cash; 251, Jack Dermid, Grant Heilman; 252, Alexander Klots, Emil Javorsky; 253, Jack Dermid, Popperfoto; 255, Novosti; 259, Weldon King; 260, Leonard Lee Rue III; 261, Holton; 263, Rolf Blomberg from Full Hand; 264, Willis Peterson; 266, W. Puchalski; 267, G. G. Collins; 269, Grant Heilman; 273, FAO Photo; 274, Stackhouse; 277, Philip Hyde; 278, Don Worth; 279, Joe Stacey; 281, J. Allan Cash, Alan B. Stebbing from Black Star, Hal Harrison; 282, Lewis Walker; 283, Willis Peterson; 284, Robert Leatherman; 285, Lorus and Margery Milne; 291, Michael Wotton; 293, E. M. and N. F. Walker; 294, J. Allan Cash; 297, Arthur Christiansen; 299, Fred Bruemmer; 300, Official U.S. Coast Guard Photograph, Suinot from Jacana; 301, Alwin Pedersen, Fred Bruemmer; 302, Fred Bruemmer; 304, Robert T. Orr from Photophile; 305, Ann and Myron Sutton; 307, B. Korobeinikon from Novosti; 310, T. Iwago; 313, Lorus and Margery Milne; 318, John H. Gerard from National Audubon Society, Laurence Pringle from National Audubon Society, Walter Dawn from National Audubon Society, David Mohrhardt from National Audubon Society; 319, John H. Gerard from National Audubon Society; 320, Lorus and Margery Milne (both); 321, A. W. Ambler from National Audubon Society; 324, Lynwood Chace from National Audubon Society; 325, Karl H. Maslowski from National Audubon Society; 329, Lorus and Margery Milne; 330, Harald Schultz.

Diagrams and maps drawn by Elaine Jones.

574.5
MIL
MILNE, LORUS
  The arena of life

DATE DUE			
			ALESCO